C000150122

Astronomy an
in the Islan

The New Edinburgh Islamic Surveys
Series Editor: Carole Hillenbrand

www.edinburghuniversitypress.com

Astronomy and Astrology in the Islamic World

Stephen P. Blake

EDINBURGH
University Press

For Meg – with love as always

Edinburgh University Press is one of the leading university presses in the UK. We publish academic books and journals in our selected subject areas across the humanities and social sciences, combining cutting-edge scholarship with high editorial and production values to produce academic works of lasting importance. For more information visit our website: www.edinburghuniversitypress.com

Edinburgh University Press Ltd
The Tun – Holyrood Road
12 (2f) Jackson's Entry
Edinburgh EH8 8PJ

Typeset in 11/13pt Monotype Baskerville by
Servis Filmsetting Ltd, Stockport, Cheshire,
and printed and bound in Great Britain by
CPI Group (UK) Ltd, Croydon CR0 4YY

A CIP record for this book is available from the British Library

ISBN 978 0 7486 4910 5 (hardback)
ISBN 978 0 7486 4909 9 (paperback)
ISBN 978 0 7486 4911 2 (webready PDF)
ISBN 978 1 4744 1319 0 (epub)

Published with the support of the Edinburgh University Scholarly Publishing Initiatives Fund.

Contents

Colour plates

Preface

To compose a readable, nontechnical account of astronomy and astrology in the Muslim world is challenging. The topic is scientific (dependent on arcane mathematical and physical theories and concepts), the period is long (covering nearly 1,000 years), the geography is extensive (stretching from India in the East to Spain in the West), and the context is crucial. To make sense of the Islamic era (from the middle of the eighth century CE until the middle of the sixteenth century), the narrative must begin three millennia before (with the Egyptians) and continue through the century following (with Copernicus, Kepler, and Newton).

Up to now, the treatments that are available fall into one of two categories. On the one hand, the books and articles by historians of Islamic science are admirably complete and sophisticated – full of formulas, diagrams, and explanations. Men like E. S. Kennedy, David Pingree, and David King have studied the Arabic treatises, carefully laying out the contributions of Islamic astronomers and mathematicians. Other historians, George Saliba, Seyyed Hossein Nasr, and Julio Samso, for example, have written longer less mathematical studies of particular topics or regions – cosmology, planetary theory, or Andalusia. And Aydin Sayili has compiled an exhaustive history of the observatory in the Muslim world.

The second category is the general history. The best of these, like the surveys of John David North, are useful for their context, situating the Islamic achievement in the larger framework of astronomy worldwide, but they are necessarily brief. Muslim astronomers and mathematicians are given no more than a chapter or two – only the most illustrious mentioned at all.

This book, on the other hand, offers a different perspective. It aims, in the first place, to be complete, covering the entire range of the nearly one thousand years of Islamic astronomy and astrology – from the first translations and compositions in al-Ma'mun's House of Wisdom in mid-eighth century Baghdad to the observatories and treatises of Raja Jai Singh in mid-eighteenth century Shahjahanabad (Old Delhi). It also aspires to be inclusive – covering not only the famous and illustrious (Nasir al-Din Tusi, al-Biruni, and Ulugh Beg) but the comparatively neglected as well – the Ottoman Taqi al-Din, the Mughal Jai Singh, and the many other scholars and scientists from Spain, Egypt, Iran, Iraq, and India who played important roles in the development of both the science and the pseudoscience. To situate the individual astronomers and astrologers in the context of their own societies is another theme, to see them in the social, cultural, religious, and scientific milieu from which they sprang. Finally, there is a good deal of comparison across regions and through time. How, for example, did the *Alfonsine Tables* of Cardoba (1270) compare to the *Zij-i Ilkhani* of Nasir al-Din Tusi (1272)? And what impact did the work of the earlier astronomers have on the observational programs, instruments, and theories of the latter?

The second feature of this essay is the effort to place the Islamic millennium in the larger history of astronomy and astrology in Western Eurasia – from the Egyptians in the third millennium BCE to Copernicus, Kepler, and Newton in the sixteenth and seventeenth centuries CE. The first chapter traces the antecedents of the Muslim era – the Egyptians, Babylonians, Mesopotamians, Greeks, Indians, and Iranians. In order to pinpoint the Islamic achievement it is important to distinguish what the early Muslim scientists took from their forbearers. And the last chapter indicates the impact Muslim astronomers and mathematicians had on the revolutionary breakthroughs of the sixteenth and seventeenth century Europeans. Although the Muslim scientists questioned the details of the Ptolemaic, Earth-centred system they had inherited, pointing out contradictions and offering corrections, they never offered a thoroughgoing alternative. It was not until the heliocentric hypothesis of Nicholas Copernicus that a completely new model was put forward. But it is important to highlight the role that Muslim scientists played in this transformation. Until the late-sixteenth century, it was the astronomers and mathematicians of the Islamic world (not the European) who stood at the forefront of the science, and it was their insights and discoveries that paved the way for the grand revolution that followed.

Since, in the Muslim world, astrology was tightly intertwined with astronomy, both must be analysed in any overall treatment of the heavenly sciences. Throughout the Islamic era and beyond the bond between the two was indissoluble – the needs of the pseudoscience driving the observations, formulas, and hypotheses of the science. Ptolemy set the agenda – composing in the second century of the Common Era the definitive texts for both astronomy and astrology – and the connection between the two remained close throughout the Islamic and early European periods. In both worlds the best astronomers and mathematicians were often the most popular astrologers; many scientists made their living by casting horoscopes, predicting and interpreting eclipses and comets, and determining the best time for marriages, battles, and journeys. A scientist as brilliant as Johannes Kepler, for example, once described himself as a Lutheran astrologer.

Although historians of science – both those mentioned above and many others as well – provided the raw material for this essay, there has been no attempt here to emulate their command of mathematics, astronomy, and physics. Rather, the effort has been to simplify the scientific side of the story while expanding its social, cultural, and comparative aspects. Leaving out most of the complicated formulas and difficult diagrams, a general sense of the individual contribution has been given in nontechnical terms. Although important terms and concepts are described in the text and a glossary of instruments is appended, some difficulties must inevitably remain. Either the narrative has been simplified too much – incompletely or inaccurately describing the mathematics or physics – or it has not been simplified enough – leaving too many technical and scientific terms unexplained. The hope is that a middle ground between these two extremes has been reached so that the splendid achievement of the Islamic scientists is presented in a way that is both complete and comprehensible.

CHAPTER ONE

From Egypt to Islam

In order to understand the work of Muslim astronomers it is necessary to return to the beginning – not just to the early Egyptian and Babylonian stargazers – but to the first beginning of all: the Big Bang. It is only in the last fifty years or so that cosmologists and astrophysicists have come to an agreement about the origins of the universe. About fourteen billion years ago a hot, dense, primordial mix exploded, and the rapid expansion that followed led to the vast cosmic configuration that we find today. Cosmologists estimate that our universe contains approximately one hundred billion galaxies – our Milky Way is one. And each galaxy in turn contains about one hundred billion stars – our Sun, the centre of our solar system, is one. Formed about 4.6 billion years ago, our solar system includes planets, moons, dwarf planets, comets, meteors, and an asteroid belt. The Sun is the principal component of the solar system, containing 99.9 per cent of its mass and dominating its gravitational field. Eight planets orbit the Sun. In order they are Mercury, Venus, Earth, Mars, Jupiter, Saturn, Uranus, and Neptune. Until 2006 Pluto, the farthest from the Sun, had also been listed as a planet but it has now been reclassified as a dwarf planet. The planets differ in size, mass, composition, temperature, and distance from the Sun. The inner planets (Mercury, Venus, Earth, and Mars) are relatively small, composed mostly of rock and have few or no moons. The outer planets (Jupiter, Saturn, Uranus, Neptune, and the dwarf Pluto) are massive, mostly gaseous, and have rings and moons. The Earth is the densest planet and Jupiter is the largest.

From the dawn of human history man has gazed at the heavens and wondered at the spectacle: the rising and setting of the Sun, the shifting phases of the Moon, the movements of the planets, and the patterns of the night-time stars. The Sun was probably the first celestial body to be studied. Its apparent annual motion on the ecliptic (the apparent path of the Sun on the celestial sphere) was always eastward, but not perfectly uniform. The length of the tropical year was 365¼ mean solar days, but the Sun's apparent angular speed eastward on the ecliptic varied, from a maximum of sixty-one minutes per mean solar day on 1 January to a minimum of fifty-seven minutes per mean solar day on 4 July. As a result, the length of the seasons also varied: winter (eighty-nine days), spring (ninety-three), summer (ninety-four), and autumn (eighty-nine). Because the Earth rotated on its axis at an angle of about 23.5 degrees as it travelled around the Sun, the hours of daylight also varied. In the northern hemisphere when the

Sun crossed the celestial equator at the spring (20 or 21 March) and autumn (20 or 21 September) equinoxes, the twenty-four hours were equally divided – twelve hours of light and twelve hours of darkness. As spring advanced to summer the Sun moved progressively north and the hours of daylight increased, until at the summer solstice (20 or 21 June) it was light for fifteen hours or more. From the summer solstice the Sun reversed direction and moved south until at the winter solstice (20 or 21 December) it reached its maximum southern point, leaving only about nine and a half hours of daylight. From the earliest period of settled civilisation the connection of these movements to the agricultural year was well known, and the important solar positions were tracked, recorded, and marked. Many prehistoric monuments seem to have been oriented toward the midsummer and midwinter risings and settings of the Sun: Stonehenge is one example.[1]

At many of these early monuments the Moon was also an object of scrutiny, its more complex movements captured in the various orientations and alignments. Revolving around the Earth in approximately 29½ days, the Moon changed shape as it went – swinging from new (almost invisible) to full (completely round) and back again. Like the Sun it also shifted its position of rising and setting, moving eastward along the ecliptic in a non-uniform motion that varied from about eleven degrees to about fifteen degrees a day, averaging thirteen. The orbit of the Moon was inclined to the ecliptic at an angle of 5 degrees, 8 minutes, and its path showed a number of irregularities or perturbations, due mostly to the gravitational force of the Sun. The size of the largest irregularity was first estimated by the early Greek astronomer, Hipparchus. Eclipses of the Sun and Moon occurred periodically. A lunar eclipse took place when the Earth passed between the Moon and the Sun, and the Earth's shadow obscured the Moon. It happened at night and was relatively frequent. A solar eclipse, on the other hand, occurred when the Moon passed between the Earth and the Sun blocking all or a portion of the Sun. Happening during the day, it was much rarer and could only be seen in a restricted area and for a short time.[2]

The motions of the planets and all other bodies in the solar system were governed by Newton's law of universal gravitation: two bodies attract one another with a force that is directly proportional to the product of their masses and inversely proportional to the square of the distance between them. Since the Sun was by far the most massive object in the solar system, it had the strongest gravitational force, and the orbits of the planets depended, for the most part, on their individual mass and their distance from the Sun. According to Kepler's first law of planetary motion, the planets circled the Sun in elliptical orbits – the Sun at one focus of the ellipse. In such an orbit a planet's distance from the Sun and its speed varied, moving faster the closer it got to the Sun, slowing down as it moved farther away. The distance from the Sun and period of the five visible planets were: Mercury (36 million miles away, 88 days), Venus (67 million miles,

226 days), Earth (93 million miles, 365 days), Mars (142 million miles, 686 days), Jupiter (778 million miles, 12 years), and Saturn (887 million miles, 29.5 years).

From the perspective of the earthbound observer the motions of the heavenly bodies were difficult to decipher. The Sun moved with a nearly constant angular speed always toward the east, and the Moon's motion was also eastward but more complicated – faster and more variable. The apparent paths of the planets, however, were much more complex – reversing direction, looping and zig-zagging, speeding up, slowing down, and even stopping. This weird activity perplexed the ancients. To the Egyptians, Mars was 'one who travels backward', and to the Babylonian the planets were 'wild sheep'. And the English word itself ('planet') came from the Greek 'wanderer'. Although, from the Earth-centred point of view the motions of the planets were bizarre, from the Sun-centred, Newtonian perspective they were completely intelligible. The Earth circled the Sun more quickly than the superior planets – Mars, Jupiter, and Saturn. Like a faster car on a freeway, the Earth overtook and passed these slower planets. As the Earth sped past Mars, for example, the Red Planet appeared first to stop and then to drop back or recede, finally resuming its original direction but making a loop in the process. For the inferior planets (Mercury and Venus), orbiting more quickly than Earth, the process was reversed. They moved rapidly west of the Sun during their retrograde motion, then slowly overtook the Sun after their motion became direct. They appeared as morning stars when on the west side of the Sun and as evening stars when on the east. To describe planetary motion astronomers adopted a special vocabulary. The longer-lasting eastward motion was called progression or direct motion, and the shorter westward motion was retrogression or retrograde motion. The time and place when direct motion changed to retrograde was called the stationary point or simply the station of the planet.[3]

To the naked eye the Sun, Moon, and planets moved against a fixed background of innumerable twinkling stars. Of the one hundred billion or so stars in the Milky Way, some of the closest or brightest were visible from the Earth, and they seemed to form clusters or constellations. Over the centuries these constellations were identified and given names. In the northern hemisphere some of the most prominent were Ursa Major (or the Big Dipper), Ursa Minor (or the Little Dipper), Cassiopeia, Andromeda, Cancer, Capricorn, Gemini, and Orion. No one knows how many of the other stars in the Milky Way have planets or how many of those other planets (if there are any) are capable of sustaining life.[4]

Although in northern Europe the evidence for early astronomy was primarily archaeological, in Egypt there were written records going back to the third millennium BCE. The great god of the early Egyptians was Ra, the Sun god, and the first pyramids (c. 2800 BCE) were oriented toward the Sun. Many of the rituals of the early pharaohs were also Sun-based. Some myths, however, were lunar, centred on the full Moon, and were associated with fertility because the thirty-day cycle of the Moon was the same length as the female menstrual cycle. For the early Egyptians, though, the most important astronomical event seems to have been the first sighting of the star Sirius, the brightest in the sky. Its heliacal rising or first appearance, which took place in mid-July, coincided with the beginning of the Nile flood – which, irrigating the entire valley, was the focal event of the agricultural year. Reconciling the three interlocking temporal systems – solar, stellar, and lunar – was an extremely difficult task that occupied Egyptian astronomers for many centuries. The festival of Sirius followed the solar year of 365 ¼ days but the lunar year (12 months of 29.5 days) was only 354 days. As a result, from about the middle of the third century BCE the Egyptians created one of the oldest calendrical systems – the lunisolar. They added an extra or intercalary month to the calendar every three years or so, bringing the lunar and solar years into rough equivalence. With time, this scheme was further refined: the year was set at 365 days and the 12 months were divided into three ten-day 'weeks'. The problem of the extra quarter day was at first ignored.[5]

Although the Egyptians paid close attention to the heavens and venerated certain celestial bodies, they did not keep any systematic records of cosmic activities – movements of planets, eclipses, or the like. The Babylonian dynasties of Mesopotamia, in contrast, kept meticulous records and developed sophisticated mathematical techniques for analysing the movements of the heavenly bodies. Mesopotamian astronomy and astrology can be divided into four periods. The first centred on the reign of the great Babylonian ruler Hammurabi (c. 1792–1750 BCE). During this period the Babylonians took over the cuneiform script and number system of the Sumerians, who had ruled the Mesopotamian city states for centuries. Using a mixture of phonetic spelling and Sumerian ideograms, the Babylonians developed their own script and left a large collection of documents on fragments of dried clay. The Sumerian number system was crucial because it employed place-value notation. That is, like our present-day decimal system, but unlike the numeral system of the Romans, a number's significance depended on its position. In the number 222, for example, the individual digits represent different values depending on their location in the sequence. The Sumerian and Babylonian systems, however, differed from the decimal in that they were based on a scale of sixty rather than ten. Although deciphering the meaning of a number in the sexagesimal (base 60) system can be difficult for the uninitiated, the legacy of the Babylonian system has persisted to the present-day: in our reckoning of time – hours, minutes, and seconds – and measuring of angles – degrees, minutes, seconds. To help ease the burden of calculation the Babylonians also devised a computational system employing number tables – multiplication tables and tables for reciprocals, squares, and square roots.[6]

Under Hammurabi the calendar was unified and Babylonian names were given to the months. Intercalary rules were also adopted, deciding whether a month would have twenty-nine or thirty days and whether a year would have thirteen rather than twelve months. The Babylonians also had an interest in what would come to be known as astrology. Many of their omens were concerned with the planet Venus, and there are tables going back nearly to the reign of Hammurabi giving the time and place of the rising and setting of the planet. These tables were particularly important because they showed that the Babylonians realised that Venus' motions shifted periodically.[7]

The second period of Babylonian history was the Assyrian (1000–612 BCE). During this period the early astronomers compiled a record of the rising and setting of some thirty-six stars. Inscribed on a series of clay tablets, these lists contained the names of constellations and star patterns and their related omens. They were also reports of lunar eclipses, rules for calculating the rising and setting of the Moon, and data on the height of shadows cast by a one-foot gnomon.[8]

The third era of Babylonian history encompassed a time of independence (612–539 BCE) followed by a period of Persian rule (539–331 BCE). During this era omens gave way to a new kind of divination based on the horoscope (see Glossary), and, as a result, systematic observation began of all the planets – not just of Venus. These records, known as astronomical diaries, began as early 652 BCE and listed planetary positions, lunar eclipses, solar halos, earthquakes, epidemics, water levels, market prices, and weather changes. The Babylonians calculated the 18-year or 223-month period over which the cycle of solar and lunar eclipses repeated itself. They determined the equivalent periods of the lunisolar calendar: nineteen solar years equalled 235 lunar months, and they computed the periods of the planets – Venus, for example, made five circuits of the Sun and eight returns to the same place in the stars over an eight-year period. By the early fifth century BCE the Babylonians had come up with the beginnings of a coordinate system. They divided the sky into the familiar twelve signs of the zodiac, each encompassing a thirty-degree segment of the great circle. Each of the zodiacal signs was named after one of the familiar constellations or star groups. In order, beginning with the vernal equinox, they were: Aries, Taurus, Gemini, Cancer, Leo, Virgo, Libra, Scorpio, Sagittarius, Capricorn, Aquarius, and Pisces.[9]

The last period of Babylonian astronomy was the Seleucid (331–247 BCE). The establishment of a system of celestial coordinates was of great importance for mathematical astronomy, essential for an accurate and systematic analysis of planetary motion. The motives of the Babylonians were partly scientific but also partly religious and astrological. Of particular interest was Zoroastrianism, at this period the dominant religion of Iran. Its teachings located the home of the human soul in the heavens and asserted that the heavens influenced terrestrial activities and events. To practice horoscopic astrology, however, there needed to be a method for tracking planetary movement. The oldest known cuneiform horoscope (dated to 410 BCE) depended on clay tablets that were early versions of an ephemerides – tables listing the positions of the Sun, Moon, and planets at specific intervals (monthly, weekly, or daily). The Babylonians developed two separate methods for calculating and predicting the odd, seemingly erratic movements of the heavenly bodies – both of which were mathematical rather than geometrical.[10]

Babylonian astronomical tablets showed evidence of two processes: first, the creation of theories attempting to explain planetary motion, and, second, rules specifying how to predict astronomical phenomena using these theories. Most of the surviving tables represented the results of the second process – that is, the compilation of ephemerides. Greek astronomy seems to have developed in roughly the same way but at a much later date, beginning in about the second century BCE. By this time the Greeks had adopted a different approach to astronomy – geometrical rather than strictly mathematical. They modelled the heavens on a sphere containing the Sun, Moon, planets, and stars and explained heavenly variation as the result of the rotation of the celestial sphere. Using this model the Greeks were able to explain the workings of the universe on a rational basis.[11]

The discovery that the Earth was a sphere has been ascribed to Parmenides in the sixth century BCE, and he is thought to have proved that the Moon was illuminated by the Sun. But Greek astronomy of the fifth century BCE was primarily concerned with meteorological phenomenon – clouds, winds, thunder, lightning, rainbows, and so on. The planetary theory of Eudoxus (c. 408–355 BCE) was the exception. Interested in music and arithmetic from an early age, Eudoxus travelled to Athens to study with Plato. He visited Egypt a few years later and calculated an eight-year calendar cycle. It was, however, his contributions to arithmetic, geometry, and astronomy that made his reputation. Although Eudoxus was responsible for major advances in number theory and for some of the finest sections in Euclid's *Elements of Geometry*, it was his planetary theory that attracted the greatest interest. His planetary model was constructed of transparent spherical shells, one within the other, concentric with a fixed spherical Earth. The shells rotated uniformly at different speeds. With this system Eudoxus was able to describe the direct and retrograde motions of both the Sun and the Moon with three spheres each. For the planets four spheres were required and for the background of fixed stars one – bringing the total number of spheres to twenty seven. Eudoxus himself made no attempt to connect the various spheres with one other. To him the thickness and size of the spheres, their suspension and their order – all were immaterial. He was not concerned with numerical accuracy either. It seems likely that Eudoxus regarded his system as an abstract, theoretical construct, a testing ground for geometrical theorems. Despite these shortcomings, however, the planetary theories of Eudoxus were the first serious attempt to explain the retrograde motions of the celestial bodies in a rational, non-mythological manner.[12]

Although Plato (427–347 BCE), the giant of Greek philosophy, had been the teacher of Eudoxus and later of Aristotle, he contributed little to the development of Greek astronomy. His primary influence was his theory of the universe. According to Plato, the world was closed and finite and everything was interior – the seven planets, the sublunar region, and the Earth. The cosmos as a whole had a soul and so did the other heavenly bodies. The fixed stars were spherical, rotating balls of cold fire, each possessed of a soul that made it forever follow its twenty-four hour circular course about the Earth, the centre of the cosmos. The planets were also made of cold fire, free from gravity, mass, and other sublunar imperfections. Each had a soul that caused it to naturally follow its complicated path through the heavens. The souls (and movements) of the heavenly bodies had been given to them by the Great Artificer or Prime Mover. According to Plato (and later Eudoxus and Aristotle) the seven planets orbited the Earth in the following order: Moon, Sun, Venus, Mercury, Mars, Jupiter, and Saturn.[13]

Aristotle (384–22 BCE), the most influential ancient philosopher of the sciences, defended the Platonic theory of the universe, representing the planets and stars as eternal substances in unchanging motion. He took the theories of Eudoxus and others and turned an abstract set of geometrical concepts into a unified system of natural philosophy that held sway for two thousand years. Aristotle studied under Plato in Athens and later had Alexander the Great as a pupil. His writings were extensive and highly systematic and covered a large part of human knowledge. His most important work on astronomy and cosmology was *On the Heavens*. In it he depicted a celestial sphere with a spherical Earth at its centre. Aristotle had two kinds of motion: celestial motion (which applied to the massless celestial bodies composed of ether) and terrestrial motion (which applied to the ponderous sublunar bodies composed of the four elements: earth, air, fire, and water). For celestial bodies the natural tendency was to move in circles; for sublunar materials it was to move in straight lines. For heavy bodies, natural movement was toward the centre of the universe; for light bodies it was away from the centre. Since bodies on the Earth fell straight down, and fire ascended vertically, the Earth could not be rotating. If it were, the individual particles would have a natural circular motion.

Aristotle's planetary system was that of Eudoxus with the addition of twenty-two more crystalline spheres. Aristotle's perfect celestial realm was unique, ungenerated, and eternal. In his *Metaphysics*, Aristotle spelled out the technical details of his system. In addition to the spheres which reproduced the motions of the heavenly bodies, Aristotle also postulated a series of counteracting spheres which were needed to neutralise the effects of the spheres above them – Jupiter for Saturn, Mars for Jupiter, and so on. In his model there were fifty-five spheres, a grand total that included the direct and counteracting spheres of each planet. Aristotle's mechanistic model was a universe of spherical shells. Motions were no longer postulated as though they were items in a geometry book but were now explained in terms of a physics of motion, of cause and effect. The first sphere of all, the first heaven, exhibited perpetual circular motion, which it transmitted to all lower spheres. The mover of the first heaven, however, was itself unmoved and eternal, the Unmoved Mover or First Cause.[14]

Apollonius of Perga (c. 262–190 BCE), a Greek mathematician, was a key figure in the development of the epicycle concept. Although the theory was expanded and refined by Ptolemy several centuries later, Apollonius worked through the details of using a second, smaller circle (epicycle) rotating around a larger original circle (deferent) as a way of modelling a planet's movement. Apollonius, however, did not seem to be interested in actual observation, and the first Greek astronomer to apply mathematical methods to geometrical astronomical theory was Hipparchus (fl. 150–125 BCE).

Hipparchus was one of the most original and creative of Greek astronomers. He put astronomy on the path that eventually led to the synthesis of Ptolemy three centuries later. Hipparchus fundamentally changed the role of observation in Greek astronomy by making a series of observations himself and by insisting on the importance of numerical accuracy and precision. He played a large role in making Babylonian observations available to Greek astronomers and showed how to use comparisons between old and new observations to reveal astronomical changes too slow to be detected within a single lifetime. He was the first astronomer to systematically employ the eccentric (a point near the Earth around which the heavenly bodies rotate) and the epicycle to represent the motions of the Sun and the Moon. Hipparchus also invented or greatly advanced trigonometry, radically improving the methods of numerically computing the sides and angles of plane and spherical geometrical figures. He wrote a treatise on chords (a chord is a line joining two points on a circle) and drew up a simple table of chords (similar to a table of sines). He worked out problems of spherical geometry by translating them into problems involving circles and triangles on a plane. This form of stereographic projection (a projection that pictures a sphere as a plane) became important in the early development of astronomical instruments, especially the astrolabe (see Glossary). It also made possible the mapping of stars on a plane surface and the creation of lines representing the local horizon, the meridian, and other relevant coordinates.[15]

To represent the motion of the Sun Hipparchus employed the epicycle and eccentric of Appollinus. He discovered the length of the solar year and the variability of the Sun's progress along the ecliptic. He predicted the time of eclipses to within an hour and their terrestrial latitude. For the Moon he assumed a simple epicyclic model similar to that he had used for the Sun. His theory was based on Babylonian and Alexandrian lunar eclipse observations. Hipparchus improved on Aristarchus' estimate for the Moon's distance from the Earth and discovered the second inequality in the Moon's motion (evection), determining its maximum amount to be 1¼ degrees (the modern measurement is 1 degree,

16 minutes). He also measured the inclination of the lunar orbit to be five degrees eight minutes. Hipparchus, however, left no detailed theory of planetary motion. He was hindered, Ptolemy later explained, by the paucity of accurate observations.[16]

After his discovery of a new star in c. 134 BCE, Hipparchus began to catalogue the fixed stars. He determined the positions of more than 1,000 stars and introduced the concept of stellar magnitudes. Comparing his own observations with those of the Alexandrian astronomers Timocharis and Aristyllus more than 150 years earlier, Hipparchus realised that the celestial longitudes had all increased by about the same amount, while celestial latitudes had remained about the same. This was most apparent for the stars near the ecliptic, such as Spica, which showed an increase in longitude of 2 degrees over 150 years or an average increase of about 48 seconds per year. He thought that this slow westward drift of the stars, or the precession of the equinoxes, was parallel to the ecliptic and common to all stars. Although there was some evidence that Hipparchus had originally proposed a precession rate fairly close to the now accepted value (50 seconds per year), he was later content to state a lower limit of 1 degree per century (36 seconds per year). This slower rate was adopted by Ptolemy.[17]

After the conquest of Alexander the Great (356–323 BCE) in 332 BCE, Egypt was ruled by Ptolemy I Soter, one of Alexander's generals. Ptolemy's successors moved the Egyptian capital from Memphis to Alexandria and founded two important scholarly institutions (the Museum and the Library) that served to introduce knowledge from Mesopotamia and the East into the Greek and Mediterranean worlds. In 30 BCE the future Roman Emperor Augustus defeated Marc Anthony and deposed Queen Cleopatra, Anthony's lover and the last of the Ptolemies, and Egypt became a province of the Roman Empire.[18]

Claudius Ptolemy (c. 100–170 CE.), the most famous Eurasian astronomer and mathematician before Nicholas Copernicus, was born and lived in Alexandria under Roman rule. His name suggests his heritage: Claudius, a Roman name, shows that he was a Roman citizen, and Ptolemy, a Greek name, reveals his ethnicity. His fame rests on three works: the *Geography*, a survey of the Greco-Roman world; the *Almagest*, on astronomy; and the *Tetrabiblos*, on astrology.

The *Almagest* (*Greatest*) is the major surviving text on ancient astronomy. Its title suggests its status. In the original Greek it was *The Mathematical Compilation*. Later it became the *The Great* [or *Greatest*] *Compilation* and finally ended up in Arabic as *al-Majïsti* or *The Greatest*. In Latin (translated from the Greek in 1160 and from the Arabic in 1175) it became the *Almagest*. One of the most influential scientific texts of all time, its geocentric model of the universe held sway for more than 1,400 years, from its completion in the mid-second century until *On the Revolution of the Celestial Spheres* of Copernicus in the mid-sixteenth century.[19]

The *Almagest* comprised thirteen books. Book One contained an outline of Aristotelian cosmology. The celestial realm was spherical. The Earth was also spherical and rested motionless at the centre of the universe. The heavenly bodies were spherical and circled the Earth in the following order: Moon, Mercury, Venus, Sun, Mars, Jupiter, Saturn, and, finally, a sphere of fixed stars. Book One also described the cardinal circles and points on the celestial sphere. In an introductory mathematical section Ptolemy explained how Menelaus' theorem – which dealt with triangles in plane geometry – could be generalised to apply to great circles in spherical geometry. There was an account of chords and a table of chords to three sexagesimal places (the equivalent of five-place logarithmic tables). Ptolemy also calculated the obliquity of the ecliptic, that is the angle between the ecliptic and the celestial equator, and arrived at a value of 23 degrees, 51 minutes, 20 seconds. This was a relatively poor result (the modern value is about 23 degrees, 40 minutes).[20]

Book Two covered the daily motions of the heavens: rising and setting of celestial bodies, length of daylight, determination of latitude by observing the semidiurnal arcs of the Sun, points at which the Sun was vertical, and the shadows of the gnomon at the equinoxes and solstices.

Book Three dealt with the length of the year and the motion of the Sun. Ptolemy accepted Hipparchus' solar theory and his figure for the year – 365¼ days minus ¹⁄₃₀₀. He included tables that allowed a rapid calculation of the two angles needed to settle the Sun's position. Extended and refined, these techniques were also used to calculate the more complicated movements of the planets. His solar theory was a simple eccentric model: one table calculated the mean motion of the Sun on the deferent circle and the other the angle of eccentricity. An additional equation (the equation of time) was an angle to be added or subtracted, enabling the astronomer to correct the mean position of the Sun, thereby arriving at its true position – that is, what an observer would see when he looked at the sky. For chronology, Ptolemy adopted the era of the Babylonian king Nebuchadnezzar, which began 26 February 747 BCE.

Books Four and Five covered the motion of the Moon, lunar parallax, the motion of the lunar apogee, and the sizes and distances of the Sun and Moon relative to the Earth. With the Moon as with the Sun, Ptolemy began with Hipparchus but here he improved greatly on the earlier theory. He introduced a uniform motion in the geo-centred deferent that tracked the Moon's motion more closely – to an accuracy of about 10 seconds, a small quantity in the astronomy of his day. And he more fully elucidated and calculated evection (the modification of the lunar orbit due to the gravitational effect of the Sun). Unfortunately, his model suffered from the enormous discrepancy it suggested in the distance of the Earth from the Moon, implying that the apparent diameter of the Moon varied by about a factor of two during a single revolution. Patently false, but apparently overlooked by Ptolemy, this flaw bedevilled Islamic astronomers for centuries. Book Five contained an account of the construction and use of astronomical instruments, including armillary spheres (see Glossary). It also discussed the parallax (difference in apparent position of an object when viewed along two different sight lines) of the Sun and Moon. Although the lunar parallax was fairly well determined, Ptolemy could not measure the position of the Sun with sufficient accuracy to tell anything about its parallax or distance.

Book Six dealt with solar and lunar eclipses. Although Ptolemy did not add much to Hipparchus, he did explain how the observation and timing of eclipses at two different locations would enable an observer to determine differences in longitude. This was Ptolemy's recommended method, but for ancient geographers the difficulty of simultaneous observations rendered the method more theoretical than practical.

Books Seven and Eight contained a star catalogue, listing the longitudes, latitudes, and magnitudes of 1,022 stars in thirty-eight constellations. He rated the stars in six classes – one the brightest and six the faintest. This catalogue served as the framework for all other work in the field until the seventeenth century. Ptolemy's system (both the positions and magnitudes) was based on that of Hipparchus (although his is no longer extant). He also followed Hipparchus' theory of precession (the motion of the eighth sphere) and used it to update his stellar locations.

Books Nine to Eleven provided the longitudes of the planets – both inferior (Mercury and Venus) and superior (Mars, Jupiter, and Saturn). Two different epicycle arrangements were needed for these two groups of planets. Mercury had additional difficulties and was treated in Book Nine by itself. For the planets Ptolemy had much less material from his predecessors than he had had for the Sun and the Moon. From Hipparchus he had the concept of the epicycle and some information about planetary periods. With this data he was able to construct tables of mean motions. In the Greek models of planetary motion the Sun played an important role. For the inferior planets (Mercury and Venus) the mean Sun was the centre of the deferent and the epicycle, but for the superior planets (Mars, Jupiter, and Saturn) another point (the equant) was added as the centre of the epicycle. Ptolemy introduced this new concept so as to accurately track and predict planetary positions. But his introduction of the equant broke the traditional Aristotelian principle of uniform circular motion, and this inconsistency troubled later Islamic astronomers. Ptolemy wanted not only to account for the changing motions of the planets – retrograde as well as direct – but he also wanted to make it easy to calculate their positions at any time – past, present, or future. To accomplish this he devised a series of rules (tables of special equations) that allowed the skilled practitioner to correct the tabulated mean motions, thereby arriving at the true position of the planet. Although Ptolemy's system is no longer regarded as correct, his prediction mechanisms were quite successful in describing the apparent motion of each body. For each planet the relative size of the epicycle and the deferent were determined with exceptional accuracy.

After finishing the *Almagest*, Ptolemy drafted two shorter astronomical tracts. In the *Planetary Hypotheses* he went beyond the longer treatise to present a physical realisation of the universe as a set of nested spheres, transforming his geometrical models into three dimensional spheres. This was a more sophisticated version of Aristotelian cosmology and was based on the assumption that there were no empty spaces in the universe. The circles and epicycles of the successive planets could not overlap those above, and thus the scale of the entire universe was now fixed – from the inner most circle of the Moon up to the circles and epicycles of Saturn. Since Ptolemy had a figure for the distance of the Moon from the Earth he could now estimate the distances of the other planets. This scheme was elaborated by Islamic astronomers and passed on to the scholars and poets of medieval Europe – seen most famously in the nine circles of hell in Dante's 'Inferno'.

In the *Handy Tables* Ptolemy picked from the *Almagest* the tables needed for astronomical calculation and republished them, prefacing the volume with an introduction explaining how the calculations were to be made. His aim was to make his work more accessible to the practicing astronomer/astrologer, and the *Handy Tables* became the model for the later Islamic astronomical treatises (sing., *zīj*).[21]

In addition to his treatise on astronomy, Ptolemy also wrote the most important early work on astrology – the *Tetrabiblos* (*Four Books*).[22] Although the geocentric model in the *Almagest* was later superseded by the heliocentric system of Copernicus, Ptolemy's work on astrology has never really been supplanted, remaining influential not only in the universities and courts of medieval and early-modern Europe but in the newspaper columns and internet sites of today. As begun by the Greeks in the first century BCE and perfected by Ptolemy in the second century CE, astrology had four main branches. Genethlialogy (the science of births) focused on celestial configurations at the time of birth. After constructing a horoscope or birth chart, the astrologer was able (it was believed) to predict the person's character and eventual fate. Catarchic (or beginning) astrology determined the proper moment to launch an activity. It was the opposite of genethlialogy. Instead of deducing an outcome from a particular horoscope, the astrologer specified the heavenly configuration that would most likely lead to success. Interrogatory astrology, the third branch, offered answers to various questions – based on the individual's horoscope. The fourth branch was historical astrology. Developed in Sassanid Iran and dependent on Iranian cosmological theories, historical or conjunction astrology pinpointed important turning points (birth of prophet or ruler, birth or fall of a dynasty, natural cataclysm) on the basis of planetary conjunctions (apparent passing of two planets as seen from the Earth).[23]

Ptolemy wrote the *Tetrabiblos* sometime between 139 and 161, after he had completed the *Almagest*. In Book One he differentiated between two kinds of astronomical study – the first (astronomy proper) discovered the movements of the heavenly bodies and the second (astrology) examined the changes which these movements brought about. The two topics were complementary, and Ptolemy offered several justifications for the latter. As the Sun influenced the seasonal and daily cycles of nature and the Moon affected the tides and other natural rhythms, so too the stars and planets affected meteorological and natural patterns. If the knowledge of celestial cycles could help to predict weather and its effects on plants and animals, why could not an astronomer:

> with respect to an individual man, perceive the general quality of his temperament from the ambient [surrounding environment] at the time of his birth . . . and predict occasional events, by the fact that such and such an ambient is attuned to such and such a temperament and is favorable to prosperity, while another is not so attuned and conduces to injury . . .[24]

Ptolemy also argued that astrological prediction was natural and beneficial. In the matter of fate versus free will he took a balanced position. Although an individual could not escape the greater cycles of change – for example, a man with a favourable horoscope might still die in times of war – many other events were not so greatly determined, and an individual, if properly warned, might be able to avoid disaster. An astrologer, like a physician, could recognise beforehand which events or ailments were inevitable or fatal and which were contingent or treatable and could be prevented or treated.

In Book One Ptolemy related the planets to certain humoral qualities, dividing them into pairs of opposites. They might be benefic (warming or moistening) or malefic (cooling or drying), masculine (drying) or feminine (moistening), active and diurnal (suited to the day and aligned to the Sun) or passive and nocturnal (suited to the night and aligned with the Moon). Mars, for example, was associated with destruction because it was dry and cold; Jupiter was temperate and fertilising because of its warmth and humidity. Ptolemy adopted the Chaldean personification of the planets. Jupiter and Venus were friendly while Saturn and Mars were hostile. Saturn, the farthest planet from Earth, ruled Saturday and caused those born under it to be petty, malicious, solitary, deceitful, harsh, lazy, and unhappy. Jupiter, nearest Saturn, governed Thursday and brought to those born under it love, friendship, abundance, justice, honor, security, and freedom. Ptolemy also took over the Babylonian division of the Zodiac. The twelve signs were divided into four triplicities: Fire – composed of Aries, Leo, and Sagittarius; Earth – Taurus, Virgo, and Capricorn; Air – Gemini, Libra, and Aquarius; and Water – Cancer, Scorpio, and Pisces.[25]

In Book Two Ptolemy took up the topic of earthly or mundane astrology. He offered a comprehensive review of eclipses, comets, wars, epidemics, natural disasters, ethnic stereotypes, and weather patterns. He described the genetic differences between peoples of different climates – those near the equator were short with black skin and those farther north were lighter and taller – and argued that any astrological assessment must rest on the knowledge of an individual's ethnic and national background. Given his interest in geography, it is not surprising to find him assigning astrological significance to the various countries of the inhabited world. Britain and Spain, for example, were ruled by Jupiter and Mars. The horoscope of a city (at the time of its founding) or of a ruler (at the time of his coronation) could be used to establish the characteristics and experiences of the city or rule. Book Two also dealt with the topic of eclipses – whether they foretold beneficial or destructive outcomes for nations or individuals. Meteorological matters were also important and the stars and planets affected weather patterns in predictable ways.

Books Three and Four explored genethlialogical astrology or the interpretation of individual birth horoscopes. The chart yielded three kinds of prediction: of generic qualities established prior to birth (family and parental influences), of qualities at birth (sex of child and birth defects), and of post-natal aspects (length of life, quality of mind, illnesses, marriage, children, and worldly success.) Ptolemy also discussed how to predict psychological outcomes or the quality of the soul from the birth horoscope. Book Four took up predictions of material, marital, and professional success and the topics of children, friends, enemies, and death. The book ended with a discussion of the seven ages of man. Each age was associated with one of the planets: age 0–3 with the Moon, 4–14 with Mercury, 15–22 with Venus, 23–41 with the Sun, 42–56 with Mars, 57–68 with Jupiter, and 69–death with Saturn. It was impossible to properly interpret an astrological chart without taking into consideration the age of the particular individual.[26]

Although the principal influence on Islamic astronomy and astrology was Greek, India and Iran also made significant contributions to the work of the early Muslim astronomers. The Indian and Iranian scientific traditions were themselves complex, both had absorbed the teaching of the *Almagest* and the *Tetrabiblos*, and so the indigenous astronomy in both civilisations was an intricate mix. In India astronomical beliefs and practices could be traced back to the middle of the second millennium BCE. In the *Rigveda* (c. 1500–1000 BCE), the oldest of the Hindu sacred texts, astral gods were mentioned along with various time periods – half months of fourteen to fifteen *tithis* or lunar days (the bright half was the fifteen days from new moon to full and the dark half the fifteen days from full moon to new), months of thirty days, and *yugas* or eras of great length. Because sacrifice was the principal ritual of the *Vedas*, time reckoning was of the utmost importance. The priest must chant the exact words and perform the specified actions at the precise moment. An early text advised the Vedic astronomer:

The person having correct knowledge of the movements of the sun, moon and other planets, accrues dharma which will take care of his future world; artha which will ensure his prosperity in this world; and fame that will perpetuate his memory. But a bad astronomer who misleads people by his (incorrect) calculations will surely have to go to hell and dwell there.[27]

At this early period, however, there was no clear evidence of complicated calendrical schemes or advanced mathematical techniques for calculating planetary motions.[28]

With the conquest of north-western India by the Achaemenid Dynasty (r. 558–330 BCE) in the late fifth century, Babylonian instruments and ideas reached the subcontinent: Chaldean omen literature and calendar procedures along with the gnomon. In the *Puranas*, a collection of religious texts from the second half of the first millennium BCE, the *yuga* was defined as an era of 4,320,000 years, a figure derived from a Babylonian calculation of the time it took the planets to rotate completely through the heavens.

In the first millennium CE, as trade between western India and the Roman Empire increased, Greek astronomy and astrology began to enter the subcontinent. In 149/50 a Greek astrological work, probably written in Alexandria about 100, was translated into Sanskrit prose. About one hundred years later (269/70) the astronomer Sphujidhvaja turned this prose version into a shorter poetic work entitled *Yavanajataka* (*Greek Astrology*). In the early fourth century a Greek astronomical text was translated into Sanskrit as the *Romakasiddhanta* (*Roman Astronomical Treatise*). During this period astronomy outgrew its original purpose of providing a calendar for the Vedic priests. No longer confined to the study of the Sun and Moon, Indic astronomers began to analyse the movements of the five planets – first and last visibility, duration of appearance and disappearance, distance from the Sun, retrograde motion, and movement through the various signs of the zodiac.[29]

In the succeeding centuries Indian scientists began compiling their own astronomical treatises (*siddhantas*). Undoubtedly influenced by Ptolemy's *Handy Tables* and a forerunner of the Arabic *zij* (astronomical treatise), the Indic *siddhanta* treated the usual topics: measures of time, planetary theory, arithmetic and algebraic procedures, positions of stars and planets, and astronomical instruments. In the traditional account of Indian astronomy the first of these works was the *Surya Siddhanta*. While both the author and original date of composition are unknown, the treatise is said to have been finished c. 400 CE, with the earliest recension dating to the eighth–twelfth centuries CE. Reflecting the Greek and Babylonian theories of the *Yavanajataka* and the *Romakasiddhanta*, the *Surya Siddhanta* replaced the outdated Vedic and Puranic concepts and theories. In 500 verses spread over fourteen chapters it dealt with solar and lunar eclipses, mean and true positions of the stars and planets, phases of the Moon, heliacal risings and settings, and latitudes of the stars and planets. It also treated cosmology, measures of time, and astronomical instruments. According to the treatise, a cosmic wind had initiated planetary motion, and an invisible, divine force (pushing or pulling) caused the motions of the planets to vary. For most Indic astrologers and almanac makers the *Surya Siddhanta* has remained the central text – even to the present day.[30]

The most famous early Indian astronomer was Aryabhata (c. 476–550). He completed two of the most important works in early Indic science: *Aryabhatiya*, on mathematics and astronomy; and *Arya Siddhanta*, an astronomical treatise. The *Aryabhatiya* comprised 121 Sanskrit verses divided into four chapters. Chapter One covered large units of time – *kalpa* (4.32 billion years) *maha-yuga* (4.32 million years), and *yuga* (redefined to 432,000 years). According to Aryabhata, the epoch of the present *yuga* (the Kali Yuga) was midnight 17–18 February 3102 BCE. The Sun, Moon, and planets were in conjunction at zero longitude on this date. He described the orbits and diameters of the Sun, Moon, and planets and the obliquity of the ecliptic. Book Two dealt with measurement, arithmetic and geometric progression, shadow lengths, and various algebraic equations. Book Three presented the eccentric and epicyclic methods for determining the daily positions of the Sun, Moon, and planets. It also explained the timing of the intercalary month. Book Four looked at the ecliptic and the equator, the trigonometry of the celestial sphere, and the signs of the zodiac. His definitions of sine, cosine, versine, and inverse sine influenced the early development of trigonometry. In fact, the modern terms sine and cosine were mistranscriptions of the words *jya* and *kojya* as used by Aryabhata. Transcribed in Arabic as *jiba* and *kojiba* they were misunderstood by Gerard of Cremona in eleventh century Spain as *jaib*, or fold, and translated into Latin as sine (cove or bay). Aryabhata also calculated π to five figures (3.1416) – at that time the closest approximation.

The *Arya Siddhanta*, however, has been lost. An astronomical treatise, it was known only through the commentaries of later astronomer-mathematicians such as Brahmagupta and Bhaskara. Aryabhata advanced the theory that the apparent movement of the stars was actually due to the rotation of the Earth on its axis. As was true of the similar idea of the Greek philosopher Aristarchus (c. 310–230 BCE), Aryabhata's insight was ignored because it violated common sense – to the earthbound observer there was no sensation of rapid rotation. His planetary model was basically that of Ptolemy – geocentric, planets in the same order, and moving according to a similar system of epicycles and eccentrics. Aryabhata had a theory for predicting eclipses, and he calculated close approximations of several astronomical constants. He set the agenda for the work of later Indic astronomers. They covered the same ground – presenting the material in a slightly different manner or revising and updating his constants.[31]

Brahmagupta (c. 598–665) headed the astronomical observatory in Ujjain and completed his *Brahma-Sphuta Siddhanta* in 628. A voluminous treatise of 1,008 verses, it was divided into twenty-four chapters and covered planetary motions and conjunctions, the problems of direction, space, and time, lunar and solar eclipses, and the risings and settings of the planets. Brahmagupta was a talented mathematician and the first person (apart from some Mayan scientists) to treat zero as a number in its own right. Although the Babylonians had a sexagesimal positional numeral system, they had no true placeholder or positional value – thus 2 and 20, or 3 and 30, would look the same. Brahmagupta, on the other hand, considered zero to be a number and gave rules for using it with both positive and negative numbers. With this innovation he completed the modern decimal positional numeral system (incorrectly known today as the Arabic numeral system). The nine numbers plus zero derive from the Brahmi glyphs found in Indic inscriptions of the first centuries CE.

In his *Brahma-Sphuta Siddhanta* Brahmagupta introduced a number of advances in mathematical theory and practice. In algebra he gave the solution for the general linear equation and used it in his astronomical calculations. In arithmetic he described the four operations (addition, subtraction, multiplication, and division) in the newly completed decimal numeral system. In geometry he discovered the formula for cyclic quadrilaterals, and in trigonometry he presented a sine table for calculating the longitudes of the planets. Translated into Arabic in 770 as *Zij al-Sindhind (The Indian Treatise)*, the *Brahma-Sphuta Siddhanta* introduced Indian astronomy to the Islamic world.[32]

Bhaskara (c. 600–680), a contemporary of Brahmagupta, was the foremost early commentator on Aryabhata. He finished his treatise, the *Aryabhatiyabhasya*, in 629, one year after Brahmagupta's *Brahma-Sphuta Siddhanta*. In the *Aryabhatiyabhasya* Bhaskara's aim was to defend the master, clarifying the more abstruse parts of Aryabhata's mathematical astronomy and presenting a close approximation of the sine function. In his own *siddhanta*, the *Mahabhaskariya*, he dealt with mean motions of the planets; true positions as well as velocities and applications; space, time, and directions; the computation and graphical presentation of eclipses; heliacal and diurnal rising of the Moon; and heliacal rising and conjunction of the planets. In his planetary model he employed the epicyclic and eccentric models of Ptolemy.[33]

The third Indian astronomer of the medieval period was Aryabhata II (920–1000). So named to distinguish him from his illustrious predecessor, he composed the *Maha-Siddhanta* – an astronomical treatise in verse. Of its eighteen chapters, the first twelve dealt with the typical topics of the Sanskrit *siddhanta*. The last six chapters explored the geometry, algebra, and geography necessary for calculating the longitudes of planets. He also constructed a sine table accurate to five decimal places. Because of the development of the decimal numeral system, an important aspect of Indian astronomy was its computational character. The Indic astronomers were obsessed with accuracy: They calculated planetary motions (both mean and true), time reckonings of great length, and trigonometric tables of various sorts.[34]

Indic astrology, like Indic astronomy, was a mixture of the Greek and the indigenous. As we have seen, the first Greek text translated into Sanskrit, the *Yavanajataka*, was astrological. Thus, the early Indic astrologers began with the basics of Ptolemy's *Tetrabiblos*. Indic astrologers employed the twelve signs of the zodiac, assigned planets to the seven days of the week, and used Ptolemaic theories to draw up their horoscopes. However, in astrology, as in astronomy, the Indic imprint was unmistakable. A characteristic aspect was the *nakshatra* or lunar mansion. The ecliptic was divided into twenty-seven (sometimes twenty-eight lunar mansions), each named after a prominent constellation. The names themselves came from the *Vedanta Jyotisha*, a first century BCE text, and differed from the Greek. They included Ashvini (wife of the Ashvins), Bharani (the Bearer), Kritika (old name of the Pleiades), Punarvasu (two restorers or chariots), and Hasta (the Hand). Indic astrology also featured the twin concepts of karma and samsara. According to the law of karma, a person's status or station in life was determined by his past deeds (karma), as the individual soul had cycled through countless rounds of birth and rebirth (samsara). Thus, Indic astrology, in addition to being predictive, was also retrospective. In casting a horoscope the *jyotish* (astronomer/astrologer) not only calculated the individual's nativity (position of planets at birth) but also included in his interpretation the residue from his past lives.[35]

From the relevant *siddhanta* (usually the *Surya*), the local *jyotish* produced for his clients an annual almanac (*panchangam* or *panchanga*). As its name implied (*panchangam* means five limbs), the almanac had five parts: (1) *Tithi* (lunar day), (2) *Nakshatra* (stellar mansion of the Moon), (3) *Yoga* (angular relationship between Sun and Moon), (4) *Karana* (half of a *tithi*), (5) *Var* (day of week). In addition to the five attributes, the *panchangam* contained other astrological and religious information – on festivals, birthdays, and anniversaries of holy men, eclipses, auspicious and inauspicious days, planetary positions, rising and setting of the Sun and Moon, and latitudes and longitudes of important localities.[36]

In contemporary accounts the question of the Indian influence on Islamic astronomy/astrology was answered in the story of Kanaka al-Hindi (Kanaka the Indian). According to Abu Ma'shar (787–886), the famous astronomer/astrologer of the early Islamic period, Kanaka was the foremost Indian expert at the early Abbasid court. Heir to the great Indian scientists (Arbyabhata, Brahmagupta, and Bhaskara), Kanaka (in Abu Ma'shar's telling) was the one who carried Brahmagupta's *Brahma-Sphuta Siddhanta* to Baghdad and had it translated into Arabic.

While the impulse to portray the Indian influence on Islamic astronomy/astrology as the work of a single individual is understandable, it was almost surely an invention, condensing a complex and longer-lasting process into the lifestory of a single individual. In the literature there were at least two Kanakas from Western India but Abu Ma'shar's subject was most likely the author of several Sassanid-influenced works: *Book of the Secret of Nativities* and *Book of Conjunctions*. While this Kanaka could have learned his astrology in India, the stronger likelihood is that he picked it up in Baghdad, the Abbasid (750–1258) capital. His horoscopes on Islamic history appear to have been cast by an astrologer at the court of Harun al-Rashid (786–809), following the chronology of Masha'allah (740–815), Abu Ma'shar's illustrious predecessor. Kanaka's various predictions – on early Islamic and Abbasid history – suggest that he wrote his works during the reign of al-Ma'mun (813–33), Harun's successor. Thus, while Kanaka probably was an Indian astronomer at the Abbasid court, he was almost surely not the omniscient savant who (in Abu Ma'shar's story) introduced the heavenly mysteries of the East to the Islamic world.[37]

As compared to the Indic tradition there was much less that was indigenous and original in Iranian astronomy/astrology, and so the impact of Iran on the development of the Islamic sciences was less significant. During the Achaemenid (550–330 BCE) and Parthian (247 BCE–224 CE) periods there is evidence that some concepts from Indic astrology and Babylonian mathematical astronomy had become influential but there is no systematic writing on these topics before the rule of the Sassanids (224–651 CE). During the third century CE the first two Sassanid rulers underwrote translations into Pahlavi of Greek and Sanskrit works on astronomy and astrology. In Greek these included the *Pentateuch* of Dorotheus of Sidon (c. 75) and the *Anthology* of Vattius Valens (120–75), both on astrology, and Ptolemy's *Almagest*. In Sanskrit the early translations comprised an astrological work by Farmasb and the *Romakasiddhanta*. Although the original Pahlavi versions have been lost, there remain Arabic translations of Dorotheus and of a Sassanid astrological treatise entitled *The Book of Zarathustra*. These works suggest that Iranian astronomy and astrology of the early first millennium was a complex mixture – Greek and Sanskrit concepts and theories intermingled with the indigenous Zoroastrian. In addition to these writings there was also an indigenous version of the *zij* or astronomical treatise. Three examples of the *Royal Tables (Zij-i Shahryaran)* have been found. The first, in 450, was mostly derivative, dependent on an early Sanskrit work. The second, in 556, was organised on the basis of the Indic *Zij al-Arkand*, employing the Indic methods for correcting mean longitudes. The final Pahlavi *zij* was compiled in the 630s or 640s under Yazdegird III (632–51) and used the Indic double epicycle model for planetary equations.[38]

Iranian views of the cosmos were heavily influenced by Zoroastrianism. In the *Avesta*, the collection of Zorastrian sacred writings, the Sun and Moon were represented as beneficent immortal beings. The planets, on the other hand, were wandering, unpredictable, and harmful, trying to steal the light and goodness from the two luminaries. The Zoroastrian calendar assigned patron deities to each day and month. The years too were given astrological meaning. The belief in a world year of 12,000 years, each millennium governed by a zodiacal sign, strongly reinforced indigenous doctrines of apocalypse. For the Sassanids the present age (that of Zoroaster) would witness the final defeat of the demonic forces of evil.

Thus far Iranian astronomy and astrology seem to have consisted mostly of an appropriation of Greek and Indic theories and concepts. The Iranian theory of astrological history, however, was *sui generis* and an extremely important contribution to Islamic astrology. The Sassanid astrologers combined the Zoroastrian concept of the world year with an indigenous notion about the meaning of the Jupiter–Saturn conjunctions to arrive at a theory of history. Zarathustra, the legendary founder of Zoroastrianism, was said to have been the author of a Pahlavi astrological work, called in its Arabic translation *The Book of Nativities*. It contained five books, the one on historical astrology reflecting Iranian theories of planetary conjunctions. And Book 14 of the Zoroastrian *Book of the Zodiac* contained annual political and economic predictions based on planetary conjunctions.[39]

Notes

1. John David North, *Cosmos: An Illustrated History of Astronomy and Cosmology* (Chicago: University of Chicago Press, 2008), ch. 1; Theodor Jacobsen, *Planetary Systems from the Ancient Greeks to Kepler* (Seattle: University of Washington Press, 1999), ch. 1.
2. Jacobsen, *Planetary Systems*, ch. 1.
3. Ibid.; Simon Singh, *Big Bang: The Origin of the Universe* (New York: HarperCollins, 2005), 26–9.
4. Singh, *Big Bang*, ch. 1.
5. North, *Cosmos*, ch. 2.
6. Ibid.
7. North, *Cosmos*, ch. 3.
8. Ibid.
9. Ibid.
10. Ibid.
11. North, *Cosmos*, ch. 4.
12. Ibid.; Jacobson, *Planetary Theories*, ch. 2.
13. Ibid. ch. 1.
14. Ibid.; North, *Cosmos*, ch. 4.
15. North, *Cosmos*, ch. 4; Jacobson, *Planetary Theories*, ch. 3.
16. Jacobson, *Planetary Theories*, ch. 3.
17. Ibid.
18. North, *Cosmos*, ch. 4.
19. For an overview see O. Peterson, *A Survey of the Almagest* (Odense, 1974); O. Gingerich, *The Eye of Heaven: Ptolemy, Copernicus, Kepler* (New York: American Institute of Physics, 1993).
20. Jacobson, *Planetary Theories*, ch. 4.
21. North, *Cosmos*, ch. 4.

22. Frank E. Robbins (ed.), *Ptolemy's Tetrabiblos* (Cambridge: Cambridge University Press, 1940).

23. Roger Beck, *A Brief History of Ancient Astrology* (London: Blackwell Publishing, 2007), ch. 1. David Pingree, *From Astral Omens to Astrology: From Babylon to Bikaner* (Roma: Istitute italiano per l'Africa et l'Oriente, 1997), ch. 2.

24. Available at <http://www.wikipedia.org/wiki/tetrabiblos> last accessed 3 March 2014.

25. Pingree, *From Astral Omens to Astrology*, ch. 2.

26. Derek & Julia Park, *A History of Astrology* (London: André Deutsch, 1983), 41–6; Beck, *A Brief History of Ancient Astrology*, 7.

27. Pingree, *From Astral Omens to Astrology*, 195.

28. Kripa Shankar Shukla, 'Main Characteristics and Achievements of Ancient Indian Astronomy in Historical Perspective', in G. Swarup, A. K. Bag, K. S. Shukla (eds), *History of Oriental Astronomy: Proceedings of an Internattional Astronomical Union Colloquium*, no. 91, New Delhi, India, 13–16 November 1985 (Cambridge: Cambridge University Press, 1987), 9–22; North, *Cosmos*, ch. 7.

29. Shukla, 'Main Characteristics', 9–22; Pingree, *From Astral Omens to Astrology*, ch. 3.

30. B. V. Subbarayappa (ed.), 'The Tradition of Astronomy in India, Jyotisastra' in D. P. Chattopadhyaya (ed.), *History of Science, Philosophy, and Culture in Indian Civilization* (New Delhi: Center for Studies in Civilisations, 2008), vol. IV, part 4, 20–1.

31. Shukla, 'Main Characteristics', 9–22.

32. Subbarayappa, 'Tradition of Astronomy in India', 29–31.

33. Ibid. 32–5.

34. North, *Cosmos*, ch. 7; Pingree, *From Astral Omens to Astrology*, 91–2.

35. Pingree, *From Astral Omens to Astrology*, 203–4.

36. Ibid.

37. Pingree, *From Astral Omens to Astrology*, ch. 5.

38. North, *Cosmos*, 186–8; 'Astrology and Astronomy in Iran', in *Encyclopaedia Iranica*.

39. 'Astrology and Astronomy in Iran', *Encyclopaedia Iranica*; Pingree, *From Astral Omens to Astrology*, ch. 4.

CHAPTER TWO

From Muhammad to the Seljuqs

At the birth of the prophet Muhammad (c. 570) the Middle East was divided between two great empires: to the West, the Byzantine, which ruled Anatolia, Syria, and Egypt and to the East, the Sassanid, which ruled Iraq and Iran. The religions of the two empires – Christian for the Byzantine and Zoroastrianism for the Sassanid – were monotheistic and transcendental. In addition, Jewish communities of various kinds were found throughout the area. These two empires, however, were agrarian and citied whereas Mecca in the Arabian peninsula, where Muhammad was born, was pastoral, nomadic, and pagan, although open to influences from both imperial cultures.[1]

In the sixth century of the first millennium the migratory Bedouins of the Arabian peninsula raised camels and lived in tight-knit patriarchal families and clans that shared pastures and the duties of security. Individual clans were headed by chiefs but there were no larger loyalties binding the several groups – all were independent and competitive. At this time the peninsula was unusually free of outside political and military influences, and the principal unifying force for the tribes and clans was the city of Mecca. A commercial and religious centre whose shrine (the Ka'ba) was the repository for numerous gods and idols, Mecca attracted a great many people during its annual pilgrimage. At that time a peninsula-wide truce was observed, and trade and arbitration took place alongside worship. The Quraysh, the chief merchant clan of the city, were part of an extensive trading network, carrying spices, cloth, drugs, and slaves from Africa and the Far East to the towns and cities of Syria and Iraq and beyond. By the middle of the sixth century Mecca had become one of the important caravan cities of the region.

Among the gods of the Ka'ba there was no overall hierarchy or structure. Rather, for the Arabs of the peninsula the plurality of gods reflected a fragmented view of the world – of man, society, and the cosmos as a whole. Christianity and Judaism, on the other hand, monotheistic religions which had penetrated Mecca and the surrounding area, had an entirely different philosophy. These religions preached a single god who had created a moral universe in which human beings were individually responsible for their actions. All men were equal and possessed the possibility of salvation. Mecca, with its shifting population of merchants, migrants, and tribesmen, was especially open to new visions of man and the world.

Into this period of political and religious uncertainty, amidst a populace fragmented by social, political, and economic differences, Muhammad, the prophet of Islam, was born. In 570 (the generally accepted year of his birth), Muhammad was a member of one of the weaker merchant clans. His father had died before he was born, and he was raised first by his grandfather and later by his uncle, Abu Talib. A long-distance trader, Muhammad at age twenty-five married his employer, the widow Khadija. They had four daughters and four sons, the boys all dying in infancy. By all accounts, Muhammad was an indifferent businessman, hobbled by a temperament that was more religious and mystical than commercial. In the fifteen years after his marriage he periodically retreated to the mountains outside Mecca to pray. In c. 610 he received his first revelation: 'Recite: In the name of thy Lord who created'. Over the next twenty-two years, until his death in 632, the messages from Allah continued and were collected by his followers in the Qur'an (The Recitation) – for Muslims the literal word of God. According to Islamic tradition, Muhammad was simply a conduit. The angel Gabriel spoke the holy text directly into his ear, he in turn repeating it to his hearers – thus 'Recite'. The message itself was extraordinary: a just and all-mighty God warns men of a Day of Judgment, at which time each person, on the basis of his deeds, would be consigned either to heaven or to hell. Muhammad condemned the rich and powerful for their pride, avarice, and neglect of the downtrodden.[2]

For three years Muhammad confined his preaching to a small inner circle of family and close friends. The force of his inspiration and the power of his words convinced those few that his revelation was indeed divine and his wife, cousin ('Ali), and close friend (Abu Bakr) became his first converts. In about 613 he was commanded to spread his message to the city at large. His reception, however, was disappointing, his followers coming mostly from the weak and the under-privileged – migrants, members of lesser clans, and younger sons of powerful clans. The rich and powerful Quraysh merchants met Muhammad's words with derision and disbelief, finding the idea of a single god and a day of judg-ment bizarre. Ridicule soon gave way to harassment, and an economic boycott was declared, depriving Muslims of access to the marketplace. Muhammad's message was a direct challenge to the foundations of Meccan society: the gods of the Ka'ba, the profits from the annual pilgrimage, tribal traditions, and loyal-ties to clan and chief. All were threatened by the preaching of this charismatic trouble-maker.

Muhammad's small band of converts slowly coalesced into a community. In the early years they managed to survive only through the support of his uncle and clan, the Banu Hashim. After several years, however, the movement stalled, and in 619, his numbers stuck at about one hundred and his wife and uncle dead, he decided to look further afield for support. The oasis settlement of Medina, about 210 miles to the north, was torn by a murderous rivalry between the two dominant clans. Muhammad visited, made a few converts, and was invited back. In 622, after a seventy-member delegation from both clans pledged to defend Muhammad and his followers, the prophet arrived in Medina. As the head of a new community committed to the worship of a new god, Muhammad brought peace and social solidarity to a fragmented society. Over the next ten years, until his death in 632, he introduced the ingredients that would constitute his new community and would enable it to transcend the ties of clan and tribe. Among these were the familiar five pillars: prayer (five times a day), almsgiving, pilgrimage (to Mecca), fast (for the month of Ramadan), and declaration of faith (oneness of God and prophethood of Muhammad).

More than a social and religious reformer, Muhammad was also a military leader. In order to strengthen his new community, he needed to control Mecca and the rest of the peninsula. In 624 his Bedouin warriors attacked and defeated a larger Meccan force. Muhammad's victory, seen as evidence of divine favour, bolstered his prestige and the attractiveness of his message. In 628 he negotiated a treaty with Mecca, enabling his followers to complete the first pilgrimage of the new faith. In 630, however, the treaty was broken and in the ensuing battle the Muslims emerged victorious. Returning in triumph to the city of his birth, Muhammad destroyed the idols in the courtyard of the Ka'ba and declared Mecca the holy city of Islam. He now ruled both Mecca and Medina and over the next two years the Bedouin tribes of the interior increasingly came to accept his political, military, and religious leadership, swelling the ranks of the new community.

In the first centuries of Islam, before the encounter with the Indic, Iranian, and Greek theories, a local, non-technical folk tradition of astronomy and astrology characterised the Arabian peninsula. A simple, practical body of knowledge, devoid of mathematical and geometrical rules and concepts, this tradition included star nomenclature, the rising and setting of stars, and their use as guides to watercourses and wells and as timing devices for fertilising camels. The early Arabs also possessed a detailed knowledge of the months and seasons, the apparent motions of the Sun and Moon, and the night sky as a predictor of rain. A distinctive feature of their tradition was the division of the year into thirteen-day periods (*anwa'*) defined by the risings and settings of the twenty-eight lunar mansions (daily positions of the Moon in the zodiac as it circles the Earth). Another was the custom of describing the time of day with reference to the length of a person's shadow. The Arabs also had names for the various divisions of the day and night and for the seasonal hours of daylight and night-time.[3]

In addition to its simplicity, there were two other reasons for the appeal of this tradition. In the first place, the Sun, Moon, and stars had been mentioned in the Qur'an, and a basic knowledge of their movements had become a standard part of religious training. In the second place, the Qur'an and early tradition

prescribed for each Muslim three duties, each of which was linked to the movements of the Earth, Moon, and Sun.[4]

First, a month of fasting and other festivals were defined by the lunar calendar, and the months began with the first visibility of the lunar crescent. Second
was the daily performance of five prayers – at sunset, twilight, sunrise, noon, and
afternoon. The third requirement was the performance of various rituals (call
to prayer, recitation of the Qur'an, prayer, slaughter of animals, and burial of
the dead) oriented toward Mecca. The basic conditions of these rituals – correct
time and correct orientation – were met by applying the techniques of folk
astronomy; no computation beyond simple arithmetic was required.

For the lunar rituals the actual sighting of the crescent Moon was required
or, in the case of bad weather, the counting of months of 29½ days each. For
the daily prayers it was necessary to observe shadow lengths during the day, the
rising and setting of lunar mansions at night, and the times of sunset, nightfall,
daybreak, and sunrise. The early religious specialists must have been happy to
come across arithmetical rules (from the Babylonians and the Greeks) for determining shadow lengths at midday and at the seasonal hours. They employed
these rules not only to write precise definitions for the times of daylight prayers
but also for timekeeping in general.[5]

To find the local sacred direction (or *qibla*) the early religious leaders followed
a tradition from folk astronomy. As the orientation was not just to Mecca but to
the Ka'ba itself, an ancient Meccan belief was adopted that associated the four
corners of the Ka'ba with the four regions of the surrounding world: Syria to the
North, Iraq to the East, Yemen to the South, and the lands to the West. Since
the early Muslims knew the general astronomical alignment of the Ka'ba – the
celestial phenomena associated with each corner – one way of finding the *qibla*
was to face the same direction one would face while standing outside the Ka'ba
– that is, the direction of one's region. Another rule had to do with Muhammad.
Since Medina was north of Mecca, the Prophet had always prayed facing south.
Thus, the first mosques, built within a few decades of Muhammad's death, were
oriented according to one or the other of these principles. Either they were
oriented according to the alignment of the Ka'ba or they faced due south in
imitation of the Prophet.[6]

Early Muslim authors used the expression 'science of the stars' to refer to
both astrology and astronomy. Soon, however, a distinction arose. Astrology
was defined by Abu Ma'shar, as 'the knowledge of the effects of the powers of
the stars, at a given time, as well as at a future time', and he labelled it 'science
of the decrees of the stars'. Astronomy proper became 'science of the spheres'
or '(science of the [heavenly] configurations'. In the usual classification of the
sciences, however, the two disciplines were commonly considered branches
of the same science. It was only later, during the thirteenth century, that
astronomy came to be considered a mathematical science while astrology was

shifted to the applied physical sciences, together with agriculture, medicine, and alchemy.[7]

In the first centuries of Islam folk astrology was closely linked to divination. Early sources mentioned an astrologer, named Qais ibn Nusba, who was supposed to have predicted the coming of Muhammad. He was called 'chief of his people', a title usually reserved for the learned men of the Jews, Christians, or Sabeans. However, the Muslim rejection of the ancient Arabian institution of the 'diviner priest' had a negative effect on astrology, and the Prophet was reported to have said that anyone who studied the stars studied magic.[8]

After Muhammad's death, leadership of the community passed to the prophet's closest companions – the four rightly-guided caliphs or deputies: Abu Bakr (632–4), 'Umar (634–44), 'Uthman (644–56), and 'Ali (656–61). During this thirty-year period the small group of believers in Medina was transformed into a fledgling state – with an army, a treasury, an administration, and the beginnings of a social and religious system. After the assassination of 'Ali, the prophet's cousin and son-in-law and the last of the four rightly-guided caliphs, the Islamic community came under the control of a powerful Meccan family, the Umayyads. Mu'awiya, the founder of the Umayyad Caliphate (661–750), began the process of turning his infant state – a collection of newly-conquered, loosely-governed, and far-flung communities stretching from north India in the East to southern Spain in the West – into an empire. He moved his capital from Mecca to Damascus, recruited an army of paid soldiers, and organised an administrative system. At this point no great attention was given to conversion. Jews, Christians, and Zoroastrians, as well as people of other religious beliefs, paid taxes and prospered under the peace and stability of Islamic rule. While the Umayyads left the beliefs and customs of the conquered peoples untouched, they did introduce a new medium of communication. Arabic, at first restricted to soldiers and administrators, soon became the common language of philosophers, historians, poets, mathematicians, doctors, astronomers, and astrologers.

While the Umayyads were successful in installing the framing institutions of a new empire, they were never able to completely quell the discontent that festered among the Persian-speaking believers of Iraq and Iran. Nursing a number of grievances – economic and political as well as social – they joined the opposition movement led by a descendant of 'Abbas, the prophet's uncle. After defeating the Umayyad forces in several critical battles, the Abbasids inaugurated a new dynasty.

In 762 the second ruler of the Abbasid Caliphate (750–1258), al-Mansur (754–75), transferred his capital from Damascus to Baghdad. At the centre of the new metropolis stood the imperial palace fortress, an immense walled enclosure called the Round City. In a few decades Baghdad would become one of the largest cities in the world; its rulers and citizens immortalised in the *Thousand-and-One-Nights* tales of Scheherezade. But the city was not only a place of romance, wealth, and intrigue, it was also a centre of philosophy and science. Under al-Mahdi (775–86), Harun al-Rashid (786–809), and al-Ma'mun (813–33) a wide-ranging translation movement was launched. While Muslim scholars concentrated primarily on the works of Greek philosophers and scientists, they also translated the compositions of Indian and Iranian thinkers as well. This process of mining the wisdom of the ancient world continued for several hundred years, but the high point of the activity was undoubtedly al-Ma'mun's establishment of the House of Wisdom (c. 820). There the medical works of Galen and Hippocrates, the mathematical works of Euclid, the astronomical and astrological texts of Ptolemy, the philosophical works of Aristotle, the astronomical treatises of Brahmagupta, and the astrological writings of Dorotheus were translated into Arabic. It is important to note, however, that even in these early days Islamic philosophers and scientists did not merely absorb uncritically the new material, rather they began increasingly to question what they had received, making new discoveries and advancing new theories.[9]

Among the early Islamic astronomers the Indic tradition was represented by the *Brahma-Sphuta Siddhanta* of Brahmagupta. It was translated into Arabic in c. 770 as the *Zij al-Sindhind* (*The Indian Treatise*) by Muhammad ibn Ibrahim al-Fazari (b. early eighth century, d. early ninth). Several other works with the same title were written in this early period but the most influential *Zij al-Sindhind* was the one by Abu Ja'far Muhammad ibn Musa al-Khwarizmi (b. before 800, d. after 847).

The Greek tradition, on the other hand, was embodied in Ptolemy's *Almagest*. The first translation by Yahya ibn Khalid during the late eighth century was unsatisfactory, and it was Thabit ibn Qurra (826–901) who produced the standard Arabic version during the second half of the ninth century. Once available, the *Almagest* generated a flurry of publication – summaries, commentaries, and paraphrases. Among the most important were: *anwa* texts, *haya* books, and *zij* compilations. *Anwa* texts covered the rising and setting of the lunar mansions. *Haya* books were summaries of the general principles and configurations of the heavenly bodies and were modelled after the first section of the *Almagest*. The *zij* was an astronomical treatise, like the *Handy Tables*, that enabled the astronomer to calculate the daily positions of the Sun, Moon, stars, and planets.[10]

Inspired by the Indian and Greek works the early Arabic astronomers worked out mathematical solutions to the astronomical problems of the new faith. Unlike the techniques of folk astronomy, these new methods required fresh observations with more advanced instruments and more complicated calculations. For the Moon the Baghdad astronomers developed limiting conditions for the apparent elongation of the Moon from the Sun, for the differences in the setting times of the Sun and Moon, and for the height of the Moon above the local horizon at sunset. Based on these conditions, they prepared visibility tables for specific latitudes. To better determine prayer times they applied trigonometric formulas for reckoning time from solar and stellar altitudes and compiled tables (running sometimes to tens of thousands of entries) for time-keeping from the Sun and stars at different latitudes, To establish the direction to Mecca (*qibla*), they developed cartographic, geometric, and trigonometric solutions to the problems of determining directions on a sphere. They also compiled lists of the *qibla*s for cities throughout the early Islamic world.[11]

Masha'allah ibn Athari (c. 740–815), a Persian Jew from Basra, seems to have been the first Abbasid astronomer/astrologer of note. A court astrologer and one of the men who drew up the horoscope for al-Mansur's new capital, he helped determine the exact moment (in 762) to lay the corner stone of the Round City. He remained at the Abbasid court for the next half century, com-posing works on the astrolabe and the armillary sphere and putting together a short list of astronomical definitions. His primary contribution, however, was in the field of astrology. Well-versed in the Sassanid pseudoscience, he thoroughly mastered the *Zīj-i Shahi* (the Arabic translation of the Sassanid treatise) and translated into Arabic the five-volume opus of Dorotheus of Sidon (the greatest Sassanid astronomer/astrologer). The titles of his best-known works, known mostly in Latin translations, reflect his interests: *The Great Book of Nativities*; *On Conjunctions, Peoples, and Religions*; *The Book of Dynasties and Religions*; and *The Judgments of Conjunctions and Mixtures*.[12]

The other important astronomer/astrologer of the early Abbasid world was Abu Ma'shar (787–886). The most prolific and influential Muslim astrologer, he was, in the eyes of his contemporaries, 'the teacher of the people of Islam concerning the influences of the stars'. A staple of the madrasa curriculum, he influenced orthodox and heterodox thinkers alike – from the eleventh century Isma'ilis of the Fatamid Empire to the fifteenth and sixteenth century Nuqtavis of the Safavid and Mughal empires. His authority, however, was not limited to the Islamic world, and from the twelfth century onward his writings (translated into Latin) swayed astrologers, philosophers, statesmen, and rulers in the centres of European science and politics.

The eastern Iranian city of Balkh, Abu Ma'shar's birthplace, was a centre of cultural and religious diversity – filled with Indians, Chinese, Scythians, Greco-Syrians, and Iranians who were Jews, Nestorians, Buddhists, Hindus, and Zoroastrians. A third-generation member of the Iranian, Pahlavi-oriented elite who had played a vital role in establishing the Abbasid Empire, Abu Ma'shar was a dedicated Muslim (probably a Shi'ite) who began his career in Baghdad as a Hadith (traditions of the prophet) expert. In his forty-seventh year he quarrelled with Abu Yusuf Ya'qub ibn Ishaq al-Sabbah al-Kindi (796–873), the leading Muslim philosopher and mathematician of his day. Al-Kindi even-tually convinced him to change his scholarly orientation – from the traditional sciences (Arabic grammar, Qur'anic interpretation, traditions of the prophet, and jurisprudence) to the rational sciences (logic, philosophy, mathematics, astronomy, astrology, and medicine). Employing his newly-acquired skills, Abu Ma'shar wrote a philosophical justification of astrology along with a practical guide for its everyday use.[13]

Of his forty-two works, the most important were: *Great Introduction to the Science of Astrology*; *Book of Conjunctions or Book of Religions and Dynasties*; *Book of the Revolutions of the World Years*; *Great Book of Nativities*; *Book of the Thousands*; *Treatise of the Millennia*; and *Tables of Conjunctions and Transits*.

Because most of Abu Ma'shar's writings on astronomy have been lost, his reputation rests, almost entirely, on his contributions to the field of astrology. These works fall into three categories. The first consisted of his attempt, primarily in the *Great Introduction to the Science of Astrology*, to provide a new rationale for the pseudoscience. Offered as a replacement for Ptolemy's *Tetrabiblos* (the reigning justification), Abu Ma'shar's argument relied heavily on Aristotle. According to the Greek philosopher, the higher heavenly bodies influenced not only the motions of the lower planets and stars but the activities of individuals and institutions as well. By charting the configuration of the heavenly bodies at particular times and places the astrologer was able (he claimed) to predict the destinies of individuals, institutions, and states.

The second category, and by far the most influential and far-reaching, comprised Abu Ma'shar's theory of historical astrology. While the most complete discussion is found in the *Book of Religions and Dynasties*, the *Book of the Thousands* and the *Tables of Conjunctions and Transits* also contained relevant material. Divided into eight parts and sixty-three chapters, the *Book of Religions and Dynasties* relied heavily on Zoroastrian beliefs – especially on a chronological system based, for the most part, on the conjunctions of Jupiter and Saturn. The two largest and most distant planets, Jupiter was about 500 million miles from the Sun and Saturn about 900 million (the Earth was 93 million.) As a result, it took Jupiter about twelve years to complete one revolution around the Sun and Saturn about thirty. Every twenty years or so, when the two planets appeared more or less together, a conjunction occurred. In Abu Ma'shar's theory, though, all conjunctions were not equal. Tracked across the sky through the twelve signs of the zodiac, the Jupiter-Saturn juxtapositions that marked the shifts from one triplicity (i.e., group of three signs) to the next and occurred every 240 years were the most important.[14] There were four triplicities, each named after one of the four Aristotelian elements: the first, Fire, included the zodiacal houses of Aries, Leo, and Sagittarius; the second, Earth, contained Taurus, Capricorn, and Virgo; the third, Air, had Gemini, Libra, and Aquarius; and the last, Water, included Cancer, Scorpio, and Pisces. The most significant conjunction of all, however, was the Grand Conjunction. Occurring once every 960 years, it marked the completion of a full revolution – when the Jupiter–Saturn conjunctions had cycled through all twelve signs of the zodiac and had returned to the first point of Aries, shifting from the Watery to the Fiery triplicity.[15]

In the *Book of Religions and Dynasties* Abu Ma'shar demonstrated that the great (and not so great) events of the past had been marked by conjunctions of greater (or lesser) rarity. For example, the Grand Conjunction of 3101 BCE (17–18 February) had heralded both the Biblical Great Deluge and the beginning of the Indic Kali Yuga. Other Jupiter–Saturn conjunctions were equally momentous, coinciding with the birth of the prophet Muhammad (571) and the founding of the Abbasid Empire (749).

The third and final category consisted of Abu Ma'shar's writings on genethlialogy, the science of births. In the *Great Book of Nativities* he discussed the nativity or birth horoscope – how it was cast and interpreted. According to the astrologer, the configuration of the heavenly bodies at the exact moment of birth determined or predicted the destiny of the individual – his longevity, choice of spouse, number of children, and occupation. In addition, Abu Ma'shar compiled – from the traditions of India, Iran, Greece, and Egypt – a collection of nativity maxims. His aim was to show the fundamental unity of the different interpretative schemes.

While the bulk of Abu Ma'shar's attention was directed to astrology, he did compile an astronomical handbook (*zīj*). Not the result of fresh observations, the *Treatise of the Millennia*, which aimed to restore the true astronomy of the prophetic age, drew on the three astronomical traditions of Abu Ma'shar's day. His computation of the mean motions of the planets employed the *yuga* technique of the Indic *Zīj al-Sindhind*, his prime meridian and planetary parameters were taken from the Sassanian *Zīj-i Shahi*, and his overall planetary model was Ptolemaic.[16]

Abu Ma'shar's influence, however, was not limited to his own time and place. By the first part of the twelfth century his works had been translated from Arabic (and Greek) into Latin and had become widely available in medieval Europe. In fact one scholar has argued that Abu Ma'shar was, for European thinkers, the single most important source of Aristotle's theories of nature.[17] Aristotle's writings on logic had been known much earlier, and he was generally recognised as the 'master of logic'. But John of Seville's translation in 1133 of Abu Ma'shar's *Great Introduction to the Science of Astrology* introduced the Greek's scientific theories to European scholars – Albert the Great, for example, and Adelard of Bath. Thereafter Aristotle became known as the first natural scientist, the 'master of those who know'. Abu Ma'shar's astrological interpretations of history – especially in the *Book of Religions and Dynasties* – were also widely influential. Roger Bacon, Pierre d'Ailly, and Pico della Mirandola all explained, in greater or lesser detail, the historical significance of the Jupiter–Saturn conjunction cycle. Finally, Abu Ma'shar's *Great Book of Nativities*, translated into Latin in the early thirteenth century, had a decisive impact on the work of medieval and early-modern European astronomers and astrologers.

The hundred or so years between Abu Ma'shar and Abu Rayhan Muhammad ibn al-Biruni (973–1048) witnessed the careers of several talented astrono-mer/astrologers. The most famous perhaps were al-Khwarizmi, Ahmad ibn Muhammad ibn Kathir al-Farghani (fl. 830–861), and Abu 'Abd Allah Muhammad ibn Jabi ibn Sinan al-Raqqi al-Harrani al-Sabi al-Battani (c. 858–929). Both al-Khwarizmi and al-Battani compiled astronomical treatises which greatly surpassed the compositions of their predecessors while al-Farghani wrote a descriptive, non-mathematical introduction to Ptolemaic astronomy that became widely popular in both the medieval Islamic and European worlds.[18]

The details of al-Khwarizmi's early life are obscure. He was a member of the House of Wisdom during the reign of al-Ma'mun, under whom his most impor-tant works were composed. His *Algebra*, the first treatise in Arabic on the subject and the work on which his fame in the West is primarily based, was a work of elementary practical mathematics – useful for lawyers, bankers, merchants, and builders. Its full title – *The Compendious Book on Calculation by Completion (al-jabr) and Balancing (al-muqabala)* – suggests al-Khwarizmi's approach. 'Completion' (*al-jabr*) eliminated negative quantities from the equation and 'balancing' reduced positive quantities to the same power on both sides of the equation. His treatise occasioned a great debate over his sources: Greek or Indian. Although the arguments are complex, the weight of the evidence seems to suggest that India, rather than Greece, was the major influence. His treatise, however, was only a start, and over the next several centuries algebra was further refined and devel-oped. Because of its popularity in Spain, al-Kharizmi's *Algebra* became authori-tative in western Europe as well, translated into Latin twice in the late twelfth century – by Robert of Chester and by Gerard of Cremona. Its pre-eminence is reflected in the name given to the new discipline – 'algebra' deriving from the '*al-jabr*' of al-Khwarizmi's title.[19]

Al-Khwarizmi was also responsible for introducing the concept of Hindu (or as they were misnamed, 'Arabic') numerals to the wider world. The title of his original Arabic composition, which has survived only in a later Latin translation, is uncertain but was probably *The Treatise on Calculation with Hindu Numerals*. The work explained the numbers 1–9, 0, and the place value system. It also briefly described several applications: addition, subtraction, multiplication, division, fractions, and square roots. Although popular in the Islamic world, al-Khwarizmi's treatise had an even greater impact in the West, sparking a renewed interest in the principles and techniques of arithmetic. It also gave rise to a new term: 'algorithm', from 'Algorismus'. The Latin form of al-Khwarizmi's name designated the step-by-step procedure by which a mathematical problem was solved.

Before turning to the *Zij al-Sindhind* it is important to take a brief look at al-Khwarizmi's other compositions. Like many astronomers, he compiled a geography. He listed the longitudes and latitudes of selected towns and cities and drew a map of the world. He wrote a history and a short treatise on the

Jewish calendar and era. Although his chronicle was mentioned by contemporary historians, no copy has survived. It may, however, have employed some of the historical chronology of his contemporary Abu Ma'shar. He also wrote two short works on the astrolabe – one on its construction and the other on its use.

The title of al-Khwarizmi's famous treatise is misleading. Not unique to him, *Zij al-Sindhind* first appeared as the name given to the Arabic translation of Brahmagupta's *Brahma-Sphuta Siddhanta*. The Indian treatise was the first work to employ zero as a number and to use algebra to solve astronomical problems. It was a revelation to the astronomer/astrologers of Baghdad, motivating them to compose works which, because they were all based to some degree on Brahmagupta's original, were also entitled *Zij al-Sindhind*.

It was, however, al-Khwarizmi's version which came to dominate the field. Partly, this was because of its quality – his treatise exhibiting an unprecedented grasp of algebra and the new numbering system – and, partly, it was because it happened to be the one taken up by Maslama ibn Ahmad al-Majriti (d. 1007) in tenth-century Cordoba. It was the first Arabic astronomical treatise to survive in anything like its entirety – although only in an early twelfth-century Latin translation based mostly on al-Majriti's adaptation. From the commentaries of earlier astronomers, however, we have some idea of al-Khwarizmi's original. His sine tables used Indian values rather than Ptolemaic and his epoch was the Yazdegird (16 June 632) rather than the Hijra (14 July 622). Although the form of most of his tables was Ptolemaic, the basic parameters derived from Hindu astronomy. Also, in al-Khwarizmi's treatise the motions and positions of the heavenly bodies as well as the equations and procedures were all taken from Brahmagupta. Al-Majriti's adaption, also entitled *Zij al-Sindhind*, followed al-Khwarizmi quite closely; the only substantial changes were the substitution of the longitude of Cordoba for that of Baghdad and the employment of the Hijra era rather than the Yazdegird.[20]

Al-Farghani was younger than al-Khwarizmi but he also worked in the House of Wisdom under al-Ma'mun. A skilled engineer, he constructed in 861 a new Nilometer – a structure used to measure the clarity and depth of the Nile River during its annual flood. It was, however, as an astronomer that he was best known. His introduction to the Ptolemaic system, *Elements of Astronomy on the Celestial Motions*, was for centuries the standard introduction – in both its Arabic and Latin versions. It was a clear, concise, and comprehensive account of the *Almagest*, descriptive and non-mathematical, the perfect text for students and non-specialists alike. Al-Farghani's *Elements* contained thirty chapters, following closely the organisation of Ptolemy's famous treatise. Chapter One was given over to chronology – describing the Hijra and Yazdegird eras and calendars along with those of the Romans and Syrians. Chapters Two–Five laid out the basic facts of the Ptolemaic universe: the centrality of the Earth and the spherical nature and motions of the heavenly bodies. Chapter Five presented Ptolemy's calculation of the obliquity of the elliptic and Chapters Six to Nine described the lands and cities of the seven climes. Chapters Ten to Eighteen analysed the movements of the Sun, Moon, and five planets. Chapter Nineteen was a star catalogue and chapters Twenty to Thirty-Two treated planetary distances and lunar and solar eclipses.[21]

Al-Battani's treatise – *The Treatise* or *The Sabian Treatise* (c. 901) – was, like so many others, an effort to correct the errors of his predecessors – Ptolemy in particular. Its significance, however, was due to the fact that al-Battani was able to make a great number of fresh observations and to update and adjust many of the basic astronomical parameters. In fact, he recorded his first observation in 877 and his last in 918 – a period of more than forty years. Because of his commitment to observational accuracy and his mathematical genius, his *zij* was a major influence on a many later astronomers – from al-Zarqali in Islamic Spain to Regiomontanus, Tycho Brahe, and Copernicus in the Latin West.

Unlike al-Farghani's *Elements*, al-Battani's *zij* did not closely follow Ptolemy's *Almagest*. Al-Battani's intent was more practical than theoretical and, instead of beginning his work with the basic foundational questions (as had Ptolemy and al-Farqani), he began with practical definitions: the division of the celestial sphere into the signs of the zodiac and the multiplication and division of sexagesimal fractions. In Chapter Three he presented his theory of trigonometric functions, employing the sine and cosine instead of the chord but not using either tangents or cotangents. Chapter Four contained his calculation, based on his own observations, of the obliquity of the ecliptic – 23 degrees 35 minutes – more than 16 minutes lower and more accurate than Ptolemy's 23 degrees 51 minutes. He also noted the variability over time of this important constant. Chapters Five to Twenty-Six treated a number of problems (mostly astrological) in spherical astronomy. Chapters Twenty-Seven to Thirty-One took up Ptolemy's theories of solar, lunar, and planetary motion, and chapters Thirty-Two to Forty-Eight discussed eras, calendars, and the rules for translating dates from one era and calendar to another. (Chapters Thirty-Nine and Forty dealt with the lunar parallax and the Moon's distance from the Earth – necessary for computing eclipses). Chapters Forty-Nine to Fifty-Five considered the chief problems in astrology. And the last two chapters discussed instruments – the sundial (see Glossary) and a special kind of armillary sphere.

Al-Battani corrected a number of Ptolemy's constants: he discovered the variability in the length of the four seasons due to the change over time in the Sun's apogee (closest point to the Earth in its apparent path) and eccentricity (variation of its apparent elliptical path from the circular) – confirming earlier calculations that the motion of the apogee was one degree in sixty-six years; he arrived at a more accurate estimate of the apparent diameters of the Sun and Moon; he improved the calculation of the Moon's mean motion in latitude and derived a more elegant method for computing the magnitude of lunar eclipses; and he improved Ptolemy's estimate for the length of the tropical year. Al-Battani's catalogue of fixed stars was less comprehensive than Ptolemy's (489 versus 1022), but in it he corrected the positions and magnitudes of his predecessor.[22]

Although al-Biruni is better known today for his encyclopaedic multi-volume description of India, in his own time he was famed as an astronomer/astrologer, consulted by the rich and powerful. By his time, however, Baghdad was no longer the undisputed political and intellectual centre of the Islamic world. Cairo, in the west, and Bukhara and Ghazni, in the east, had begun to challenge the primacy of the Abbasid capital, and al-Biruni's career, unlike Masha'allah's or Abu Ma'shar's, reflected this geographical and political decentralisation.[23] He was born in a suburb (*birun*) of Kath, the capital of the Central Asian state of Khwarizm, and spent his first twenty-five years mastering both branches of Islamic knowledge: the traditional and the rational. A talented linguist, he learned Arabic, Persian, Turkish, Sanskrit, Hebrew, and Syriac. Leaving Kath in the wake of political and military disturbances, he lived for the next ten years in Bukhara and Gurgan, finally returning in c. 1009 to Kath and the court of the Khwarizm Shah Ma'mun ibn Ma'mun (1009–17). In 1017 Mahmud of Ghazni (997–1030) defeated Ma'mun and annexed his lands to the recently-founded Ghaznavid state (975–1187). A widely-circulated but apocryphal anecdote identified the *causus belli* as Mahmud's demand that Ma'mun send the two famous scholars (al-Biruni and Ibn Sina (980–1037)) from Kath to Ghazni. While al-Biruni passed the remaining thirty years of his life as the chief astrologer of the Ghaznavid court, Ibn Sina migrated west, residing mostly in Isfahan and other Iranian cities.

Al-Biruni, like Ibn Sina, was extraordinarily productive. The best modern estimate puts the total number of his compositions at about 150, at least 80 per cent of which have been lost. While he is best known today for *India*, some ninety-five of his full-length works (about 65 per cent) were devoted to astronomy, astrology, and related subjects. Between 1017 and 1030 he travelled extensively in India. He learned Sanskrit, read deeply in Indic religious, philosophical, and scientific literature, and interviewed many Brahmin priests and intellectuals, translating Ptolemy's *Almagest* into Sanskrit and Patanjali's *Yogasutras* into Arabic. His *India* was a philosophico-religious treatise (an account of Hindu beliefs and creeds), on the one hand, combined with an anthropological survey (a description of Indic ceremonies, festivals, and sacrifices), on the other. Al-Biruni's detailed and sympathetic account of the Indian subcontinent was a monumental achievement, unmatched until the mid-nineteenth century work of the British administrator/scholars.

In addition to his magnum opus, al-Biruni completed works on a variety of other topics. After surveying the available literature, he compiled exhaustive treatments of pharmacology (listing almost 1,200 individual drugs) and mineralogy (a comprehensive catalogue of precious stones and metals). It was, however, as an astronomer/astrologer that he had his greatest impact. His first major composition – *The Chronology of Ancient Nations* – was preceded by seven lesser works: one on decimal computation, one on the astrolabe, one on astronomical observations, three on astrology, and two histories. The *Chronology* contained the kind of historical and calendrical information that would have been found in the astronomical treatises of his day – those, for example, of al-Khwarizmi or al-Battani. The topics included: the day and year, calendars and months, eras (intercalation for the lunisolar), chronological and regnal tables (of the Assyrians, Babylonians, Persians, Byzantines, Achaemenids, and Sassanids), prophets (Mazdak and Zoroaster), and feasts and festivals (of the Persians, Greeks, Jews, Christians, pre-Islamic Arabs, and Muslims).[24]

Al-Biruni's *Determination of the Coordinates of Cities* was another astronomical work. A driving force behind the geographical work of Islamic astronomers was the need to determine the relative location (latitude and longitude) of towns and cities in order to properly locate the *qibla* niche (direction toward Mecca) in newly-constructed mosques. To collect this information al-Biruni acted both as a researcher (mining earlier authors for their results) and as a practicing astronomer (making a number of observations himself and developing new ways to process and organise the data). He also included the results from his Indian work – specifying the coordinates of the towns and cities he had visited. He composed several treatises on the construction and operation of astronomical instruments. While his treatment of the sextant was short, his longer composition on the astrolabe was especially valuable – containing a construction manual, a list of tools, a catalogue of available models, and a theoretical critique. *The Shadows* was a detailed work of thirty chapters. In the three hundred years after the death of Muhammad the sundial had been the everyday clock for the mosque timekeeper (*muwaqqit*), enabling him to determine prayer times from the length and direction of the gnomon's shadow. However, the actual process was complicated, and a great deal of attention had been given to topics like the length of the day, the season of the year, the latitude of the mosque, the shape of the shadow, and the translation of all of this into accurate directions for the local muezzin (prayer caller). Al-Biruni also included a section on the Indian sundial.[25]

Al-Biruni's primary contribution, however, to the field of astronomy was the *Masudic Canon*. Dedicated to Mahmud's son and successor Mas'ud (1030–9), the *Canon* won the astronomer a reward and a pension. Although the work superficially resembled an astronomical treatise, it was (like the *Almagest*) concerned as much with the theoretical derivation of astronomical parameters as it was with the construction of tables for the computation of planetary positions. The *Canon* was divided into eleven chapters. Chapters One and Two were introductory, covering regnal and chronological tables, the spherical character of the Earth, units of time, and calendars. Chapters Three and Four dealt with plane and spherical trigonometry. Al-Biruni's contribution was to define π arithmetically, as the ratio of two numbers, rather than geometrically, as the ratio of two shapes. Chapter Five (taken mostly from his *Determination of the Coordinates of Cities*) gave the geographical coordinates of many urban centres. Chapters Six and Seven dealt with the Sun and Moon. While the model was Ptolemaic, al-Biruni derived several parameters from his own observations. Chapter Eight took up the calculation of solar and lunar eclipses. Chapter Nine, on the fixed stars, included a new catalogue – containing 1029 entries rather than Ptolemy's 1022. Chapter Ten was on the planets and included tables for calculating longitude, latitude, visibility, distance, and diameter. The concluding chapter dealt with mansions, transits, rays, and the historical astrology of Abu Ma'shar.[26]

Despite al-Biruni's training as a scientist and philosopher, he was, for much of his life, a practicing astrologer, and two of his compositions, although often overlooked in the secondary literature, dealt explicitly with the pseudoscience. The first and shorter was on transits (a technique for interpreting horoscopes), and a specialised topic of interest only to practicing professionals. The second, however, was longer and more accessible. Entitled the *Book of Instruction in the Elements of the Art of Astrology*, it was written for one of the ladies of the Khwarizm court and took the form of a beginner's manual. Organised in a question-and-answer format, the book contained five chapters, the first four of which were preliminary. Chapter One covered geometry (in particular spherical geometry) while Chapter Two took up mathematics: numbers, computation, and algebra. Chapter Three dealt with geography, cosmology, and astronomy, defining the technical vocabulary of the working astrologer. Chapter Four centred on the astrolabe – its operation and underlying theory. Only Chapter Five, the last, covered topics that were properly astrological. It was divided into five sections and consumed nearly half the treatise. The first section dealt with natural astrology – meteorology, earthquakes, floods, and all other natural disasters. Section Two covered political events: the outcome of battles and revolutions, the rise and fall of kingdoms. Section Three took up genethlialogy, the science of casting nativities. Section Four was on beginnings: when to found a city, start a journey, or take up an occupation. Section Five examined omens and divinations. In this work, as elsewhere, al-Biruni scattered comments on Indian traditions and practices.

Notes

1. For a discussion see Ira Lapidus, *A History of Islamic Societies* (Cambridge: Cambridge University Press, 1988), chs 1–2.
2. Ibid. ch. 2.
3. David A. King, *In Synchrony with the Heavens: Studies in Astronomical Timekeeping and Instrumentation in Medieval Islamic Civilization. Volume 1: The Call of the Muezzin* (London & Leiden: Brill, 2004), 635.
4. Ibid. 465–6.

5. Ibid.
6. Ibid. 753.
7. George Saliba, *A History of Arabic Astronomy: Planetary Theories During the Golden Age of Islam* (New York: New York University Press, 1994), 66–7.
8. Ibid.
9. Ehsan Masood, *Science and Islam* (London: Icon Books, 2009), 39–64.
10. Saliba, *A History of Arabic Astronomy*, 73–4.
11. King, *In Synchrony with the Heavens*, 466.
12. 'Masha'alah', in *Dictionary of Scientific Biography*; E. S. Kennedy & David Pingree, *The Astrological History of Masha'allah* (Cambridge, MA: Harvard University Press, 1971), v–vii, 69, 75.
13. 'Abu Ma'shar', in *Dictionary of Scientific Biography*.
14. E. S. Kennedy, 'An Astrological History based on the Career of Genghis Khan', in E. S. Kennedy (ed.), *Astronomy and Astrology in the Medieval Islamic World* (London: Ashgate Variorum, 1998), 224.
15. Ibid. 223–31.
16. 'Abu Ma'shar', *Dictionary of Scientific Biography*; David Pingree, *The Thousands of Abu Ma'shar* (London: The Warburg Institute, 1968); Keiji Yamamoto & Charles Burnett (eds & trs), *Abu Ma'sar on Historical Astrology*, 2 vols (Leiden: Brill, 2000); E. S. Kennedy, 'The World Year Concept in Islamic Astrology', in E. S. Kennedy (ed.), *Astronomy and Astrology in the Medieval Islamic World* (London: Ashgate Variorum, 1998), 351–71; E. S. Kennedy & B. L. van der Weerden, 'The World Year of the Persians', in E. S. Kennedy (ed.), *Astronomy and Astrology in the Medieval Islamic World* (London: Ashgate Variorum, 1998), 338–50.
17. Richard Lemay, *Abu Mashar and Latin Aristotelianism in the Twelfth Century, The Recoverry of Aristotle's Natural Philosophy through Iranian Astrology* (Beirut: American University of Beirut, 1962).
18. North, *Cosmos*, 196–8.
19. 'al-Khwarizmi, Abu Jafar Muhammad ibn Musa', in *Dictionary of Scientific Biography*.
20. Ibid.
21. 'al-Farghani, Ahmad ibn Muhammad ibn Kathir', in *Dictionary of Scientific Biography*.
22. 'al-Battani, Abu 'Abd Allah Muhammad ibn Jabi ibn Sinan al Raqqi al Harrani al Sabi', in *Dictionary of Scientific Biography*.
23. There is a vast literature on al-Biruni, much of it in Russian. The various encyclopedias give a comprehensive overview. See, for example, 'al-Biruni, Abu Rayhan Muhammad ibn al-Biruni', in *Dictionary of Scientific Biography*; 'al-Biruni, Abu Rayhan Muhammad ibn al-Biruni', in *Encyclopaedia Iranica*. In addition: E. Sachau (ed.), *al-Biruni's India: An Account of the Religion, Philosophy, Literature, Geography, Chronology, Astronomy, Customs, Laws, and Astrology of India* (London: Keagan Paul, 1910); E. Sachau (tr. & ed.), *The Chronology of Nations* (London: 1879); S. H. Nasr, *An Introduction to Islamic Cosmological Doctrines: Conceptions of Nature and Methods for Study by the Ikhwan al-Safa, al-Biruni, and Ibn Sina* (Binghamton: State University of New York Press, 1993); H. M. Said-Azkhan, *al-Biruni: His Times, Life, and Works* (Karachi: Hamdard Foundation, 1981); P. Chelkowski (ed.), *The Scholar and the Saint: Studies in Commemoration of Abul-Raihan al-Biruni and Jalal al-Din al-Rumi* (New York: Hagop Kevorkian Center for Near Eastern Studies, 1975).
24. Ibid.
25. Ibid.
26. Ibid.

CHAPTER THREE

The observatory in Isfahan

In the Eurasian world the astronomical observatory was an Islamic creation. Until Tycho Brahe founded his observatory at Uraniborg in 1576, fabricating astronomical instruments, carrying out a programme of observations, creating mathematical formulas and methods of computation, and composing astronomical works were all done in the Islamic world rather than in the European world. Although the first observatory with a sizeable number of large instruments and an extended programme of observation was that of the Seljuq ruler Malik Shah in Isfahan, there had been observation posts and smaller-scale observatories in the preceding centuries.

The earliest systematic observation programme in the Islamic world was conducted during the reign of the Abbasid caliph al-Ma'mun. This is not surprising for, as we have seen, he had established the House of Wisdom for the express purpose of translating into Arabic the works of the Indian and Greek scientists and philosophers. The *Almagest* and *Handy Tables* of Ptolemy were the most important of these works, and al-Ma'mun was interested in updating Ptolemy – primarily for astrological reasons, to more accurately determine the positions of the heavenly bodies. As a result, al-Ma'mun organised observations at two places – at Shammasiyya, a residential quarter within the city of Baghdad, in 828–9, and at Mt Qasiyun, in the vicinity of Damascus, in 831–2. The usual collection of small, mostly portable instruments was employed – armillary sphere, mural quadrant (see Glossary), gnomon (about sixteen feet tall), and azimuthal quadrant.

Because of the brief period devoted to observation, al-Ma'mun's results were limited. Yahya ibn Abi Mansur (d. 830), the caliph's senior astronomer, supervised the Baghdad programme and wrote the astronomical treatise that resulted – *The Well-Tested Astronomical Treatise*. Yahya's father, Abu Mansur Aban, had been an astrologer at the court of al-Mansur, and Yahya followed in his father's footsteps, becoming the personal astrologer and boon companion of the caliph and an important figure in the House of Wisdom. Although the corrections of Ptolemy in Yahya's treatise were few in number, they were not trivial. Most of the observations were of the Sun and Moon, and al-Ma'mun's astronomers calculated solar and lunar eclipses and arrived at a more accurate figure for the obliquity of the ecliptic – 23 degrees 33 minutes, more accurate than Ptolemy's 23 degrees 51 minutes. They also came up with more accurate figures for the precession of the equinoxes (about 1 degree in 66 years) and for the maximum solar equation (1 degree 59 minutes) – both improvements on Ptolemy. Yahya and his men also established the *qibla* for Baghdad and compiled a table with the longitude and latitude of twenty-four fixed stars.[1]

Other important contributors to the House of Wisdom were the three sons of Musa ibn Shakir, another one of al-Ma'mun's court astrologers. Known as the Banu Musa Brothers (Brothers [who were] Sons of Musa), Muhammad, Ahmad, and Hasan were precocious talents, studying in the House of Wisdom under Yahya ibn Abi Mansur. Using smaller, mostly portable instruments – armillary sphere or perhaps solstitial and equinoctial armillaries – they made a number of observations between about 847 and 869. In 847–8 they recorded observations of several stars in the constellation Ursa Major and in 868–9 they measured the maximum and minimum altitudes of the Sun – both of these sets of observations were conducted from their house in Baghdad, located near a bridge over the Tigris River. They also observed the autumnal equinox in Samarra and arranged for simultaneous observations of a lunar eclipse in Samarra and Nishapur in order to determine the difference in latitude between the two cities. The brothers published between them twelve works on astronomy. In addition to their astronomical publications, they also authored important works on geometry and mechanics – including a foundational treatise on the measurement of plane and spherical figures and a book on mechanical devices (a self-dimming lamp among them). They planned and oversaw the construction of a canal for a new city and recruited and patronised a number of talented astronomers and mathematicians – the most notable of whom was Thabit ibn Qurra. They could not, however, get along with al-Kindi and joined his persecutors during the latter part of their careers.[2]

'Abd al Rahman al-Sufi (903–86) was an Iranian astronomer who worked in Shiraz at the court of the Buyid ruler of Fars, 'Adud al-Dawla (949–82). The Buyids, a confederation of several tribes, controlled Iran and Iraq for most of the ninth and tenth centuries, and their rulers developed a passionate interest in astronomy and astrology. Like the rest of the early Muslim astronomers and astrologers, al-Sufi took as his goal the revision of Ptolemy. He used a ring with a radius of 125 centimetres to measure the obliquity of the ecliptic, and he also employed an equinoctial armillary for his other observations. His fame, however, derived from a rather unusual ambition: he decided to revise Ptolemy's star catalogue, 1,022 stars in forty-eight constellations. Al Sufi's *Book of Fixed Stars* (964) was a synthesis of the Greek and Arabic traditions. He updated Ptolemy's stellar coordinates and added new values for brightness, colour, and magnitude. He also related the Greek names for the stars and constellations to their Arabic nomenclature.[3]

Sharaf al-Dawla (982–9) followed his father to the Buyid throne of Fars, succeeding not only to his title but to his passion for astronomy as well. He founded an observatory in the garden of his palace in Baghdad. The Sharaf al-Dawla Observatory, as it came to be known, was a major advance over the short-lived institutions at Shammasiyya and Qaisiyun of al-Ma'mun. Sharaf al-Dawla explicitly stated that the observations were to cover the seven planets (the Sun and Moon plus the five visible planets). A building was constructed and a large collection of instruments, some of them sizeable, were collected. The ruler appointed a director and an elaborate ceremony inaugurated the new institution. Although solstice and equinox observations were made in 988, no information has survived of any attempts at correcting Ptolemy's work – no astronomical treatises, almanacs, star catalogues, or calculations or estimates of astronomical constants.[4]

The astronomer Abu Mahmud Hamid al-Khujandi (c. 940–1000) worked for the Buyid ruler of Ray, Fakhr al-Dawla (976–97), who controlled northern Iran and Iraq. Al-Khujandi designed and constructed the largest meridian arc or mural sextant (see Glossary) of his day – it had a radius of twenty metres. The stone arc was erected between two walls on a hill outside Ray, near modern-day Tehran. Faced with wood, it had copper sheets placed on its surface. At the top of the curved ceiling of the instrument, a hole of about 2 centimetres allowed sunlight to enter. Named *al-Suds al-Fakhri* after al-Khujandi's patron, the new instrument was much more accurate than its predecessors. On the older, smaller instruments readings were accurate to degrees and minutes only whereas on the new much larger sextant a greater degree of precision was possible – to seconds of arc. In 994 with this new instrument al-Khujandi determined a value for the obliquity of the ecliptic that was much more accurate than Ptolemy's. He also found the latitude of Ray and with his collection of instruments – an armillary sphere among others – he began a full programme of observations. He intended to observe all seven of the heaven bodies and to compile a new astronomical treatise – the *Zij al-Fakhri* (*The Astronomical Treatise of Fakhri*). The death of Fakhr al-Dawla, however, soon after the construction of the great arc, brought an untimely end to this ambitious program.[5]

The culmination of this first period in the history of Islamic astronomy was the Isfahan Observatory, designed and built by the astronomer-mathematician 'Umar ibn Ibrahim al-Nishapuri al-Khayyam (c. 1048–1131) for the Seljuq ruler, Jalal al-Din Malik Shah (1072–92). The Seljuq Turks, nomadic warriors from Central Asia, had conquered the lands of eastern Islam in the early eleventh century, establishing a state that extended from northern India in the east to eastern Anatolia in the west and from Central Asia in the north to the Persian Gulf in the south. Tugril Beg (1037–63), founder of the dynasty, sacked Ghazni in 1037 and in 1039 defeated Mas'ud I, taking over the western provinces of the Ghaznavid state. In 1055 Tugril defeated the Buyids and occupied their capital of Baghdad. Alp Arslan (1064–72), Tugril's successor, defeated the Byzantine army in 1071 and annexed most of Anatolia. The high point of the Seljuq

Empire (1037–1194), however, was the reign of Jalal al-Din Malik Shah. Under the direction of his famous vizier, Nizam al-Mulk, Malik moved his capital to Isfahan, founded a series of colleges, and reorganised his fiscal and military administration.

To get a sense of Malik Shah's reasons for founding the observatory we might take a look at a roughly contemporaneous Turkish work that offered advice to the wealthy and powerful – the *Qutadgu Bilig* (*Knowledge that Brings Happiness*). Of astronomers the author, Yusuf Khas Hajib, wrote: 'Then come the astronomers . . . They make calculations concerning the years, the months and the days. Oh powerful man, this calculation is very necessary.'[6]

And further,

> When you wish to start doing anything, it is first necessary to inquire whether the time is favorable for it or not. There are lucky as well as unlucky days and months. Inquire about them and choose the lucky ones . . . [7]

'Umar Khayyam was born the year al-Biruni died.[8] The historical 'Umar was a mathematician, astronomer, philosopher, and sometime poet. While he penned the occasional quatrain, he was definitely not the Omar Khayyam of Edward Fitzgerald's *Rubaiyat*, a mystic fatalist who enjoyed the pleasures of this world and was unsure about his fate in the next. The Omar of popular imagination was defined by his verse:

> A Book of Verses underneath the Bough
> A Jug of Wine, a Loaf of Bread – and Thou
> Beside me singing in the Wilderness –
> Oh, Wilderness were Paradise enow.

And:

> The Moving Finger writes: and, having writ,
> Moves on: nor all thy Piety nor Wit
> Shall lure it back to cancel half a line,
> Nor all thy Tears wash out a Word of It.

'Umar grew up and received his education in Nishapur. Like al-Biruni and other later Islamic astronomers, he was a polymath, mastering at an early age both the traditional and the rational sciences. In 1070 he travelled to Samarqand, where he completed his first major work – *Treatise on . . . Algebra*. A comprehensive treatment of the subject, this book included not only his own theories (the most important perhaps a classification of cubic equations) but a far-reaching account of the work of Indian mathematicians as well. In 1077 'Umar published his second major mathematical work, *Commentaries on Euclid*. After summarising the recent work on this famous text, 'Umar took up the foundations of geometry and the nature of irrational numbers. Like al-Biruni, he also wrote on philosophical topics. In 1080 he composed *Treatise on Being and Duty*, responding to the request of a high-ranking Seljuq official to give his views on God's creation and on man's duty to pray. Later in the same year and probably for the same reason he wrote *An Answer to Three Questions*. His other three philosophical works – *Treatise on the Universality of Being*, *The Light of Reason on the Subject of Universal Science*, and *Treatise on Existence* – were also completed at the behest of important courtiers. His poetry, on the other hand, seems to have been – for him as well as for his contemporaries – mainly incidental, the least important of his writings. Today, however, the situation is reversed, and more than one thousand of his Persian quatrains have been published. Nevertheless, since the authenticity of many of these verses has been questioned and since their choice and arrangement has often been haphazard, it is difficult to come to any reliable conclusion about the religious or philosophical views of their author.

In 1074, two years after his coronation, Malik Shah invited 'Umar to come to Isfahan. After appointing him chief astronomer, the sultan ordered the young man to oversee the construction and operation of a new observatory. While astronomers like al-Khwarizmi, al-Battani, and al-Biruni had made periodic observations of the stars and planets and had managed minor revisions of Ptolemy's *Almagest*, their equipment had been light and mostly portable (a sextant or astrolabe), their support had been uncertain, and their tables had remained generally out of date and inaccurate. Malik Shah's new observatory, on the other hand, was a real institution. It featured a securely funded staff (seven more astronomers in addition to 'Umar), a permanent headquarters, and a collection of large, immovable instruments (including a mural quadrant and armillary sphere)

The Seljuq ruler's goal for his new observatory was also ambitious. He wanted 'Umar and his staff to produce an entirely new *zij*, a complete and total up-dating of Ptolemy's *Almagest*. Unlike the other court officials and nobles, however, 'Umar understood the enormity of the sultan's request. A complete revision, he pointed out, would require a fresh set of planetary observations and would take at least thirty years to complete – the time it took Saturn to circle the Earth. The court astronomers (and unspoken, but more to the point, Malik Shah himself) might not live that long. Instead, 'Umar suggested that the shah authorise a reform of the Seljuq chronological system – the Zoroastrian calendar and the Yazdegird Era (epoch 632), both solar. Not only would this be of benefit to society at large but it would also ensure that the shah's name would live forever.[9]

To appreciate 'Umar's task and his achievement we must return to the early days of the Muslim community. One of the first issues facing Muhammad's successors was the establishment of a calendar, a systematic way of keeping track of the days. As we have seen, the predecessors of the Muslim astronomers – Egyptian, Babylonian, Greek, Indic, and Iranian – had all wrestled with the incompatibility of the two heavenly timepieces – the Moon and the Sun. And at one point or the other they had all adopted variations on the three calendrical schemes – the lunar, the solar, and the lunisolar.[10]

In the early centuries of the first millennium the inhabitants of the Arabian peninsula had a strictly lunar calendar. The months were defined by the phases of the Moon and were approximately 29½ days each. The lunar year, comprised of twelve lunations, was divided into two parts: four months of peace (three centred on the pilgrimage month) when raiding and fighting were prohibited and eight months in which warfare was allowed. However, because the lunar year contained about 354¼ days and the solar year of four seasons about 365¼, the pilgrimage month, which had originally been in the autumn, regressed against the seasons, making it progressively more difficult to find provisions for travelling and animals for sacrifice. As a result, in 412 the Arabs adopted a lunisolar calendar, intercalating a month every three years, placing it between Dhu al-Hijja (the month of pilgrimage) and Muharram (the first month of the year).

In 631, however, according to an obscure passage in the Qur'an, the prophet Muhammad was commanded to reform the pagan lunisolar calendar and era that he had inherited. 'The number of months in the sight of Allah is twelve – so ordained by Him the day he created the heavens and the earth . . . Verily *nasi* [the intercalation of a month] is an addition to unbelief: The Unbelievers are led to wrong thereby . . .'[11] This prohibition was later repeated by the prophet: 'Oh People, the unbelievers indulge in tampering with the calendar in order to make permissible that which Allah forbade, and to forbid that which Allah has made permissible. With Allah the months are twelve in number . . .'[12]

In addition to organising the days into weeks and months, the early Muslim leaders had to also find a convenient way of ordering and numbering the years. During the decade and a half following the prophet Muhammad's migration (*hijra*) from Mecca to Medina (in 622) his followers gave the years names rather than numbers. The second was the Year of Permission, the fifth was Congratulations on Marriage, and the year of his death (632) was Farewell. After Muhammad's death, however, 'Umar, the second caliph, realised that a more conventional chronology was needed. Thus, in 638 he established the Hijra Era. The departure of Muhammad from Mecca in 622 was chosen as the starting point because, according to tradition, the prophet's followers could not agree on the date of his birth. The beginning of the new era, however, was not the actual day of the prophet's emigration but was rather the first day of the lunar year in which it took place. Thus, 1 Muharram AH 1 was 16 July 622.[13]

Although Muhammad and the early Muslims accepted the seven-day Judeo-Christian week, the problem of finding a special day of worship was difficult. According to an early tradition, the prophet stated: 'The Jews have every seventh day a day, when they get together [for prayer] and so do Christians: therefore, let us do the same.'[14]

Eventually, Friday (sunset Thursday until sunset Friday) was chosen as the peak day, keeping it near but separate from the holy days of the other two religions. Perhaps, however, feeling the need to further differentiate themselves, the early Muslims did not make Friday a day of rest. Unlike Saturday for Jews or Sunday for Christians, Friday for Muslims was a day of ordinary activity (except for the noon prayer). To further distinguish themselves they also settled on different names for their days. Thus, Friday, the day of the special community prayer, was the 'Day of Assembly' (Yawm al-Jum'a) and Saturday, under Jewish influence, was the Sabbath (Yaum al-Sabt). The other days, however, were simply the first day (Sunday), the second day (Monday), the third day (Tuesday), the fourth day (Wednesday), and the fifth day (Thursday).

In the early years, the beginning of the month and the number of its days varied. A new month could not be declared until the first slim crescent had appeared. Soon, however, in order to simplify astronomical calculations and to establish specific dates for rituals and celebrations, Islamic astronomers adopted a schematic calendar in which the months were given a definite number of days: (1) Muharram, 30 days; (2) Safar, 29 days; (3) Rabi' I, 30 days; (4) Rabi' II, 29 days; (5) Jumada I, 30 days; (6) Jumada II, 29 days; (7) Rajab, 30 days; (8) Sha'ban, 29 days; (9) Ramadan, 30 days; (10) Shawwal, 29 days; (11) Dhu al-Qa'da, 30 days; (12) Dhu al-Hijja, 29 or 30 days. The extra day was sometimes necessary because twelve revolutions of the Moon totalled about 354¼ days. In a thirty-year cycle the additional day was added in the second, fifth, seventh, tenth, thirteenth, sixteenth, eighteenth, twenty-first, twenty-fourth, twenty-sixth, and twenty-ninth years. The names of the months were pre-Islamic and did not change but they soon lost their seasonal connotations.[15]

After the death of 'Ali (656–61), the prophet's son-in-law and the last of the four rightly-guided caliphs (or deputies), the fledgling Umayyad state emerged. Headquartered in Damascus, the Umayyads governed a rapidly expanding territory that extended from North India in the east to North Africa and the Iberian peninsula in the west. As the spoils of conquest drew to an end, the need for a regular system of taxation became increasingly urgent. In an agrarian economy, the lunar Hijra Era was seriously deficient. Taxes had to be collected soon after the annual harvest, and with a lunar era the date of collection did not fall at the same point in the seasonal cycle. Any date in the Hijra Era year – for example, 1 Muharram (New Year's Day) or 10 Ramadan – regressed at the rate of about eleven days per year against the seasons. Thus, 'Abd al-Malik (685–705), the third Umayyad ruler, decided, for administrative and fiscal reasons, to adopt a new solar era. However, this new era could not be allowed to challenge the primacy of the Hijra. Thus, the Kharaji or Taxation Era (AK), with the 365-day Zoroastrian calendar (twelve months of thirty days plus five at the end) had to be designed to track the liturgical era. In the early years of its adoption this meant that the solar era numbering sequence had to be adjusted so that it would overlap the Hijra, 68 AK matching 68 AH, and so on. But, because of the eleven day difference, after thirty-two years the two eras would diverge. Left uncorrected, this discrepancy would have introduced confusion into official documents and records. It would also have left the peasant at the mercy of unscrupulous tax collectors. After paying the taxes for the current solar year, the cultivator could be duped some months later for another payment – due because the lunar year had changed. To avoid this confusion the Kharaji Era had to be periodically recalibrated, the numbering sequence advanced. This process was called *indhilaq* (sliding) in Arabic and *tahwil* (changing) in Persian. Thus, in every Kharaji Era century the years after the thirty-second, the sixty-fourth, and the ninety-sixth were eliminated – 34 AK following 32, 66 following 64, and 98 following 96.[16]

As time passed, however, this method of remedying the deficiencies of the liturgical era proved increasingly unsatisfactory. It was difficult for administrators, tax collectors, record keepers, and historians to remember the 'sliding' required to keep the Kharaji Era synchronised with the Hijra. Thus, many of the early agrarian empires adopted one or more solar eras, with numbering sequences separate from the Hijra. In pre-Islamic Iran under both the Achaemenid (559–330 BCE) and Sassanid (224–651 CE) dynasties the solar Zoroastrian calendar named and grouped the days. Both empires, however, employed a regnal system for numbering the years, the crowning of a new ruler inaugurating a new era. Thus, a battle or a birth would be in 'the twelfth year of Darius' or 'the seventh year of Cyrus' When, however, the invading forces of 'Umar (634–44), the second caliph, defeated the Sassanid army at the battle of Nihavand in 641, the regnal system of historical accounting came to an end. Yazdegird III (632–51), the last Sassanid ruler, had come to the throne in 632 and so the era that he inaugurated did not end with his death. The Yazdegird Era employed the 365-day Zoroastrian calendar. Since, however, there was no provision for the extra quarter day, the date of the New Year (Naw Ruz) had slowly regressed. In the late Sassanid period the New Year began at the summer solstice (21 June) but during the first centuries of Islamic rule it was set at the vernal equinox (21 March). By the reign of Malik Shah in the middle of the eleventh century, however, Naw Ruz had retreated again – to late February.[17]

The Seljuq tax collectors and accountants were well aware of the difficulties the unadjusted calendar posed. Since Malik Shah's state, like most other pre-modern polities, was agrarian-based, it was critical that the calendar track the seasons. Taxes could only be collected after the crops had been harvested and, as a result, the due date must be expressed in solar rather than in lunar terms. Thus, 'Umar Khayyam's suggestion that a new calendar be created – one that would return Naw Ruz to the vernal equinox – was met with instant approval.

The new solar calendar, however, required a new solar era, and the *Tarikh-i Jalali* or Jalali Era (epoch 21 March 1079) was named after the ruler who had commissioned it. The Jalali calendar contained 365 days but included a leap year – every fourth year six days rather than five were added to the basic 360. The months were the traditional Zoroastrian: (1) Farvardin; (2) Urdibehesht; (3) Khordad; (4) Tir; (5) Mordad; (6) Shahrivar; (7) Mehr; (8) Aban; (9) Azar; (10) Dai; (11) Bahman; and (12) Isfand. Unlike the Islamic, Jewish, Christian, or Indic months, however, the Zoroastrian were not divided into weeks. Rather, each day had its own name.

In addition to managing the new observatory and overseeing the creation and introduction of the new calendar and era, 'Umar also authored two other astronomical works. His *Zij-i Malik Shahi* (*Astronomical Treatise of Malik Shah*) (1090), while bringing to mind the initial desire of the shah, never really lived up to its name. All that remains in the single manuscript that has survived are several tables of ecliptic coordinates and a catalogue of the 100 brightest stars. While the date of the work (1090) suggests that 'Umar and his men could have drawn on fifteen years of observations, the truncated nature of the treatise suggests that the astronomer was trying to placate the shah rather than attempting any substantial revision of Ptolemy.[18]

After the death in 1092 of both Malik Shah and Nizam al-Mulk, 'Umar fell into disfavour. In the political turmoil that followed he was judged disloyal (because of his position as chief astronomer). Although the new regime withdrew its support from the observatory, 'Umar remained in Isfahan, earning his living as a practicing astrologer, trading on his reputation as the chief astronomer of Malik Shah and Nizam al-Mulk. His only composition during this period (and the last of his career) was the *Naw Ruz Nama* (*History of the New Year*). Another astronomical/astrological work, this was a history of the ancient Iranian solar festival and an account of the pre-Islamic calendar which it headed. In an attempt to convince Malik Shah's successors to reconsider their decision concerning the observatory, 'Umar described the magnificent celebrations of the Achaemenid and Sassanid rulers and their lavish support of religious scholars and educational institutions. Although the *Naw Ruz Nama* failed its argumentative purpose, 'Umar remained in Isfahan for another twenty-six years. On the accession of Malik Shah's third son, Sanjar, to the throne in 1118, 'Umar finally left the capital for Marv, the new Seljuq capital, where he died some thirteen years later.

There is no doubt that the signal achievement of 'Umar and the Isfahan Observatory was the *Tarikh-i Jalali*. And while Malik Shah was undoubtedly gratified by a new solar calendar and era named after himself, he may also have been disappointed that his original goal – an updated astronomical treatise – was not realised. The *Zij-i Malik Shahi*, at least in the version that survived, was wholly inadequate. Before, however, concluding that Malik Shah's ambition for his institution was unrealised, one must look into the circumstances surrounding the compilation of a full and newly-revised treatise dedicated to Malik Shah's son and successor, Sultan Ahmad Sanjar (1118–53).

Al-Zīj al-Mu'tabar al-Sanjari (*The Esteemed Astronomical Treatise for Sanjar*) was compiled by 'Abd al-Rahman al-Khazini (fl. 1115–35), an accomplished astronomer and mathematician who also wrote treatises on mechanics and astronomical instruments. Although his treatise was dedicated to Sultan Sanjar, one authority stated that it was actually completed before his accession and the catalogued version is dated 1120, just two years after his coronation. Given these dates and the number of fresh observations, it is tempting to conclude that much of the work was done at the Isfahan Observatory after the death of Malik Shah. It is even more tempting to wonder whether 'Umar, active not only in Isfahan from 1092 (the death of Malik Shah) until 1118 (Sanjar's accession), but also available in Sanjar's new capital of Marv for thirteen more years (until his death in 1131), might not have been involved in the new treatise as well – as al-Khazini's teacher or mentor, perhaps?[19]

In any event, the *Al-Zīj al-Mu'tabar al-Sanjari* was an impressive achievement. At the beginning of one version al-Khazini stated that he had made a number of fresh observations: he had observed, calculated, and compared the positions of the five planets at conjunctions and the Sun and Moon at eclipses, and he was familiar with the works of his predecessors, especially al-Biruni, Thabit ibn Qurra, and al-Battani. The first section of the treatise, the chronological, was quite extensive – covering the Hijra, Yazdegird, Seleucid, Jewish, Soghdian, and Hindu calendars. He had compiled tables of Muslim, Zoroastrian, Christian, and Jewish feasts and festivals as well as regnal tables for the Babylonian, Achaemenid, Macedonian, Coptic, Sassanid, Umayyad, Abbasid, Byzantine, North African, Buyid, and Seljuq dynasties. There was also a list of important prophets. Tables were included for lunar mansions, for plane and spherical trigonometric functions, and for equations of time. He also measured the obliquity of the ecliptic at Isfahan.

Al-Khazini's material on mean motions was, like the rest of his treatise, exceptionally complete and precise. Basic mean motions were given for all five planets and for the Moon the rate of double elongation was given to seven or more fractional places. Following Abu Ma'shar, al-Khazini presented a good deal of astrological material on world days and years. He also included numerous planetary tables – equations and latitudes, sectors, and parallax. His visibility tables were among the most extensive ever and seem to have been worked out by himself – nothing taken from the *Almagest* or from Thabit ibn Qurra or al-Battani. In the earliest version of his *zij*, al-Khazini prepared star tables for the year 500 AH (1106–7 CE). He noted the latitude, longitude, brightness, and magnitude of forty-six stars. The treatise ended with several astrological tables, including al-Biruni's theory for the projection of rays.[20]

In addition to his astronomical treatise, al-Khazini composed works on mechanics, specific gravity, and astronomical instruments. The latter had seven parts, each devoted to a different instrument: parallactic ruler (see Glossary), dioptra for measuring apparent diameters (see Glossary), a triangular instrument, a quadrant, a reflection instrument, an astrolabe, and devices to aid the naked eye. He also composed a text on a self-rotating globe marked with stars and celestial circles. In 1131 al-Khazini drafted a revised abridgment of his *zij*. Gregory Chioniades, an Orthodox bishop who travelled to Tabriz in the 1290s, translated this version into Greek, and it became the basis for a revival of interest in astronomy among the Byzantine scholars of Constantinople.[21]

The century or so between the Isfahan Observatory and al-Khazini's *zij*, on the one hand, and the work of Nasir al-Din Tusi and his collaborators at the Maragha Observatory, on the other, marked an important turning point in Islamic astronomy. The emergence of the office of *muwaqqit* (timekeeper) in the local mosques of eastern Islam signalled a turn toward a more serious consideration of the problems of mathematical astronomy – in particular, the increasingly apparent deficiencies and contradictions in Ptolemy's two major works – the *Almagest* and the *Planetary Hypotheses*.

During the twelfth and thirteen century a new subspeciality within Islamic astronomy appeared – the science of astronomical timekeeping or *'ilm-i miqat*. Basically, this involved determining the proper times for the five daily prayers. In the earliest traditions these were defined according to astronomical criteria. The Muslim day began at sunset and the interval during which the first prayer was to be performed lasted from sunset to nightfall. The interval for the second prayer began at nightfall and lasted until daybreak. The third prayer was to be performed during the interval between daybreak and sunrise. The permitted time for the fourth prayer began when the Sun crossed the meridian and ended when the interval for the fifth prayer began; namely, when the shadow of an object equalled its meridian shadow increased by the length of the object. The interval for the fifth prayer lasted until the shadow increased again by the length of the object or until sunset.

Determination of prayer times either by observation or by computation was a relatively simple problem for a competent medieval astronomer. The midday and afternoon prayer times could be established using a gnomon, or the corresponding solar altitude could be computed and the times determined with an astrolabe or quadrant. To find the times from sunrise to midday and from midday to the beginning of the afternoon prayer were standard problems of medieval spherical astronomy. The prayers at nightfall and daybreak could also be determined either by observation or by calculation. Thus, from an astronomical or mathematical point of view ascertaining the times of prayer was a straightforward application of certain standard procedures in spherical astronomy. The *muwaqqit*s prepared mathematical tables which displayed prayer times throughout the year for a particular locality. It was convenient to have such times tabulated, either for each degree of solar longitude or for each day of the year, particularly in localities where clouds were frequent or where the local horizon was obscured by mountains or tall buildings. Even with the tables, however, some kind of instrument for measuring the passage of time was required – an astrolabe, a quadrant, or a sundial.[22]

Along with their commitment to the problems of timekeeping, the *muwaqqits* also developed an interest in the more general problems of theoretical astronomy and, in particular, of planetary theory. During the twelfth and thirteenth centuries these men begin to look more critically at Ptolemy's work. They noticed that certain of the astronomical quantities set out in the *Almagest* were inaccurate – for example, the movement of the solar apogee. Some astronomers simply corrected Ptolemy while others pointed out the error and replaced it with a more accurate one derived from observation. Ptolemy's value for the precession of the equinoxes was also found to be seriously deficient and was replaced in many of the works of these later astronomers.

A more serious problem, however, was the philosophical contradiction underlying Ptolemy's planetary theory. In the *Planetary Hypotheses* Ptolemy had argued that planetary spheres were actual physical bodies (solid spheres) not just mathematical constructs or empty spheres, as he seemed to imply in the *Almagest*. If the two treatises were meant to describe the same universe, then the mathematical hypotheses in the *Almagest* presented several problems – that of the equant being the most prominent. In the models of the planets in the *Almagest* Ptolemy had inserted a mathematical point (the equant) around which the planetary epicycles moved at a uniform speed. Although a mathematical success, the equant was an absurdity in the world of physics. A physical sphere cannot move at a uniform speed around an axis that does not pass through its centre. As these problems in Ptolemaic astronomy became more apparent, the Islamic astronomers of the period began to take increasing note and to propose solutions. This increasingly sophisticated mathematical and astronomical work occurred in two places: in the West in Cordoba and Toledo, and in the East in Maragha.[23]

Notes

1. Thomas Hockey, et al. (eds), *The Biographical Encyclopedia of Astronomers* (New York: Springer Reference, 2007), 1249–50; Aydin Sayili, *The Observatory in Islam and its Place in the General History of the Observatory*, 2nd edn (Ankara: Turk Tarih Kurumu Basimevi, 1988), 50–87; North, *Cosmos*, 209–10.
2. Sayili, *Observatory*, 92–4.
3. Ibid. 104–6.
4. Ibid. 112–18.
5. Ibid. 118–21; North, *Cosmos*, 209–10.

6. Sayili, *Observatory*, 235.

7. Ibid.

8. For a brief overview see 'al-Khayyami, 'Umar ibn Ibrahim al-Nishapuri', *Dictionary of Scientific Biography*. Also consult Edward Fitzgerald, *Rubaiyat of Omar Khayyam*, ed. Christopher Decker (Charlottesville: University of Virginia Press, 1997); S. G. Tirtha, *The Nectar of Grace: Omar Khayyam's Life and Works* (Allahabad, 1941); D. S. Kasir (tr.), *The Algebra of Omar Khayyam* (Beirut, 1972); H. J. J. Winter & W. Arafat, 'The Algebra of Omar Khayyami', *Journal of the Royal Asiatic Society of Bengal, Sci.* 16 (1950): 27–77.

9. Sayili, *Observatory*, 160–6.

10. This discussion is based in part on my recent book: *Time in Early Modern Islam: Calendar, Ceremony, and Chronology in the Safavid, Mughal, and Ottoman Empires* (Cambridge: Cambridge University Press, 2013), 107–9.

11. Qur'an 9:36–7.

12. 'Tarikh', *Encyclopaedia of Islam*, 2nd edn.

13. E. G. Richards, *Mapping Time: The Calendar and its History* (Oxford: Oxford University Press, 1999), 234.

14. Ibid. 26.

15. Frank Parise (ed.), *The Book of Calendars* (New York: Facts on File, 1982), 71.

16. The first documented example was in 242 AH / 856–7 CE during the reign of the Abbasid Caliph, al-Mutawakkil (232–47 AH / 846–62 CE). H. Taqizadeh, 'Various Eras and Calendars used in the Countries of Islam', *Bulletin of the School of Oriental and African Studies* 9 (1937–9): 903–22; 'Tarikh', *Encyclopaedia of Islam*, 2nd edn, 263.

17. 'Tarikh', *Encyclopaedia of Islam*, 2nd edn; Taqizadeh, 'Various Eras', 917.

18. 'al-Khayyami', *Dictionary of Scientific Biography*.

19. Sayili, *Observatory*, 177–8.

20. E. S. Kennedy, *A Survey of Islamic Astronomical Tables* (New York: American Philosophical Society, 1956), 159–61; Hockey, *Biographical Encyclopedia of Astronomers*, 629–30.

21. Ibid.

22. King, *In Synchrony with the Heavens*, 201, 205.

23. Saliba, *A History of Arabic Astronomy*, 73–6.

CHAPTER FOUR

Astronomy and astrology in al-Andalus

After the early centuries, the centre of Islamic astronomy moved from Baghdad and the House of Wisdom to cities and scientists both East and West. In the East al-Biruni in Ghazni and 'Umar Khayyam in Isfahan followed Masha'allah and Abu Ma'shar in Baghdad. In the far West, however, in the Islamic kingdoms of southern Spain (al-Andalus), the heavenly sciences developed in a slightly different way. And this new direction was significant because, until the sixteenth century, the principal source of European knowledge about astronomy and astrology was Islamic and the principal path of transmission was by way of al-Andalus.

In 711 Arab and Berber forces under the Umayyad general Tariq ibn Zayid defeated the Visigothic king Rodrigo at the Barbate River in southern Spain and moved north. Expanding on their initial success, the Islamic forces soon controlled the entire Iberian peninsula – their advance not finally halted until 732 at Tours in southern France. After several North African governors were unable to pacify and administer the new province, an independent Spanish-Muslim state came into being. Founded by 'Abd al-Rahman I (756–88), a grandson of the Umayyad Caliph Hasham, the Umayyad Caliphate (756–1031) in Spain had its capital in Cordoba. Over the following 250 years, especially under 'Abd al-Rahman II (822–52) and 'Abd al-Rahman III (912–61), a rich and prosperous Spanish-Islamic civilisation developed. A professional army provided security and a reorganised administration built irrigation canals and collected taxes. New forms of poetry and prose appeared, and new styles of architecture, exemplified by the great mosque at Cordoba, reflected the successful melding of the Arab East and the Christian West. The Spanish Umayyads patronised the works of Arab philosophers and scientists and underwrote the first translations of the Greek masters – Aristotle, Plato, Galen, Euclid, Hippocrates, and Ptolemy – from Arabic into Latin.

In Andalusia, however, as in the eastern Islamic world, unity and political stability were difficult to achieve. In the early eleventh century the Arab clans revolted, the Caliphs lost control of the central government, and the provincial governors declared their independence. The Caliphate was abolished and was followed by the rule of the 'party kings', a period (1030–90) characterised by independent warring states in Cordoba, Seville, Toledo, Granada, and Saragossa. While political fragmentation did not significantly affect commercial prosperity or cultural vigour, it did leave the smaller kingdoms less able to resist Christian counterattacks. In 1085 Alfonso VI (1077–1109), ruler of Castile, Leon, and Galicia, conquered Toledo, and over the next fifty years the kingdom of Aragon took back Saragossa, Huesca, Tortosa, and Lerida.

The Christian reconquest of Andalusia was viewed with alarm by the Almoravid Dynasty (1040–1147) of Morocco. In 1086 a Muslim army crossed the Straits of Gibraltar and defeated Alphonso VI. From 1090 to 1145 the North Africans reconquered the Islamic city states, governing Andalusia as an Almoravid province. Although the Almoravids were rather quickly overthrown by the Almohads (1121–1269), a revolutionary mainland movement, the situation in Islamic Spain remained unchanged – one Moroccan state replacing another. Christian resentment of Islamic rule, however, persisted, and in 1212 the combined forces of Leon, Castile, Navarre, and Aragon defeated the Almohads at a crucial battle in southern Spain. Soon after the divided Islamic city states fell again to the resurgent Christians: Cordoba (1236) and Seville (1248) to Castile and Leon, and Valencia (1238) and Murcia (1245) to the Aragonese. The Portuguese captured the western part of the peninsula and by the mid-thirteenth century only Grenada remained Muslim. Protected by a large population, a mountainous terrain, and Christian disunity, it continued independent and prosperous until its final conquest by the united forces of Castile and Aragon in 1492.

During the first centuries of the Reconquista, Muslim cultural and intellectual life remained rich and vital. Arabic poetry, influenced by the land and people of Spain, flourished while debates on Islamic mysticism, theology, and philosophy generated new answers to old questions. In architecture, the Alhambra – the magnificent palace-fortress complex built by the sultans of Granada in the mid-fourteenth century – manifested the continued vitality of Muslim life and culture. After the conquest of Toledo and Sargasso in the late eleventh and early twelfth centuries, Alfonso X (1252–84) of Castile, Leon, and Galicia had the Bible, Talmud, and Qur'an translated into Castilian, and in Toledo the Archbishop had Arabic works on astronomy, astrology, and philosophy translated into Latin – works of Ptolemy, al-Kindi, al-Farabi, and Avicenna. Between 1160 and 1187 Gerard of Cremona translated some eighty-seven works into Latin – the Qur'an, Aristotle, and Avicenna, among others. And between 1220 and 1250 Maimonides's and Averroes's interpretations of Aristotle were translated into Latin and were quoted by Thomas Aquinas. In this way, Greek philosophical and scientific thought re-entered Europe through the Arab world.[1]

In the history of Islamic astronomy and astrology, Andalusia was not so much the site of scientific advance as it was the conduit through which knowledge flowed from the Islamic East to the Christian West. Before the Muslim invasion in the early eighth century the Spanish Visigoths were unacquainted with Ptolemy and Greek science. The astronomy and astrology of pre-Islamic Spain can be discussed under two headings. The first has to do with astrology – divination, beliefs in religious prophecy, alchemy, aruspicina (divination by looking at entrails of animals), and talismans.[2] Given this background, Islamic astrology was quickly accepted. The earliest work of note was a short composition – a series of astrological predictions – by 'Abd al-Wahid ibn Ishaq al-Dabbi (788–96). It was based on a Visigothic Latin treatise, entitled *Libro de las Cruzes* (*Book of the Crosses*) in the twelfth-century Spanish translation. The *Libro* had a complicated textual history – the Latin original was lost and al-Dabbi's Arabic version was composed at the end of the eighth century. It covered rain, drought, and their consequences for prices, vegetation, and illness. The forecasting techniques were simple, based on the positions of Jupiter and Saturn in the four triplicities. In a later chapter al-Dabbi took account of solar and lunar eclipses in the Earth and water triplicities. At this period the astrologer did not need to know the exact positions of the planets. To cast a horoscope according to the *Libro* he needed only to know the signs for Saturn, Jupiter, Mars, and the Moon. The prominence of Jupiter and Saturn, of course, implied a detailed knowledge of Abu Ma'shar, but the lack of information on planetary positions suggested that astronomical tables were either not available or not understood.[3]

At the Umayyad court this simplified astrology was quickly adopted. The early caliphs had court astrologers, and during the reign of 'Abd al-Rahman II these men offered interpretations of the solar eclipse of 17 September 833 and of the massive eruption of shooting stars on 20 April to 18 May 839.[4] Later, the conjunction of Jupiter and Saturn in 1006–7 occasioned a spate of dire predictions. Some astrologers declared it a sign that the end of the caliphate was near while others predicted a change of dynasty accompanied by ruin, slaughter, and famine. And in a late version of the *Libro* the conjunction was said to foretell the end of Arab rule in Spain and its replacement by a Christian or Berber dynasty.[5]

The second aspect of pre-Islamic Spanish astronomy was the emergence of a distinctive kind of sundial. A fairly primitive version, this early instrument featured a vertical gnomon fixed in the centre of a horizontal semicircle. The hours were determined by radii which divided the circle into twelve equal parts. The Spanish sundial was related to an early Latin family of church sundials called, in medieval England, scratch dials or mass clocks.[6]

Significant development of the heavenly sciences in Islamic Spain had to wait until the second half of the tenth century under 'Abd al-Rahman III (912–61) and his successors, al-Hakam II and Hisham II. At that point a group of scholars in Cordoba, having acquired the works of the Abbasid astronomers, began to write on mathematics, astronomy, and astrology. The most important of these early Andalusian astronomers was Maslama ibn Ahmad al-Majriti (c. 950– c. 1007). Not much is actually known of the man himself (the dates of his birth and death are uncertain), but he seems to have settled in Cordoba early in life.

He studied with a famous geometrician named 'Abd al-Ghafir ibn Muhammad and was engaged in astronomical observations in 979. He likely served as court astrologer for Hisham II. Although he was responsible for introducing the famous Isma'ili text – the *Rasa'il Ikhwan al-Safa* (*Epistles of the Brethern of Purity*) – to the Umayyad court, his principal accomplishment, and that for which he is best known, was his adaptation of al-Khwarizmi's well-known *Zij al-Sindhind*. Al-Majriti's work carried the same title and followed the earlier work quite closely. His only substantial changes were to switch the standard meridian from Arin (present-day Ujjain, India) to Cordoba and to employ the Hijra era rather than the Yazdegird. He also added a number of astrological tables – concerning the projection of rays, for example. These were easier to use and more accurate than those of al-Khwarizmi. It is important to remember, however, that al-Khwarizmi's *zij* was a very early example of the genre. Although its form was that of the *Almagest*, its basic parameters were Indic – motions and positions of the planets, sine tables, and equations all from Brahmagupta rather than Ptolemy. In this respect then, al-Majriti's *zij* was already outdated. By the early eleventh century the eastern astronomers had, with the *Alamgest* and the *Planetary Hypotheses* as their guides, moved considerably beyond the *Zij al-Sindhind*.

Al-Majriti's other compositions, also adaptations, were less influential. His work on commercial arithmetic covered sales, taxes, and surveying and included guides to the relevant arithmetical, geometrical, and algebraic operations. He composed a very brief treatise on the astrolabe, revised some tables from al-Battani's *zij*, and drafted some notes on Menelaus' theorem. He also completed an Arabic translation of Ptolemy's *Planishaerium* (*Star Chart*). His influence in Andalusia, however, was due mostly to the work of a talented group of disciples. Al-Kirmani (d. 1006) carried the *Epistles of the Brethern of Purity* to Zaragoza and the northern frontier. Ibn al-Samh (d. 1035) compiled astronomical tables using Indian methods and published a two-part work (130 chapters) on the astrolabe. And Ibn al-Saffar (d. 1034) composed a work on the astrolabe. These men, and others, extended his influence throughout southern Spain, giving to the astronomy and astrology of the area an Indic tinge which it never lost.[7]

After al-Majriti, the most important Andalusian astronomer/astrologer was Abu Ishaq Ibrahim al-Zarqali (c. 1029–1100). Zarqali (the 'blue-eyed one' from 'zarqal', meaning 'blue') was part of a new school of Spanish astronomy. Al-Majriti had worked in Cordoba, but after the collapse of the Umayyad Dynasty in the middle of the eleventh century, Spanish Islam disintegrated into a collection of warring city states: Cordoba, Seville, Toledo, Granada, and Saragossa. And during the last half of the eleventh century Toledo, primarily because of al-Zarqali's leadership, replaced Cordoba as the centre of Andalusian astronomy and astrology.

Unlike the other astronomers of Andalusia, al-Zarqali was entirely self-taught. He was born into a family of Visigothic converts to Islam and was

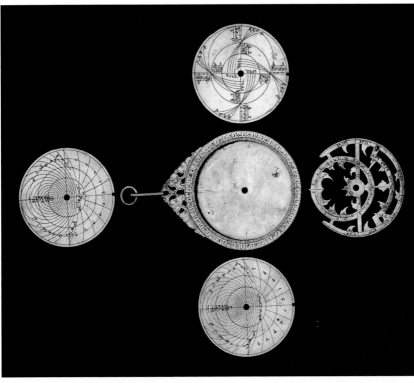

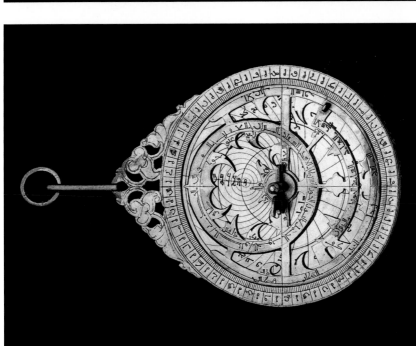

Plate 1 Thirteenth-century astrolabe and its component parts, made by Ibn Shawka al-Baghdadi. © National Maritime Museum, Greenwich, London.

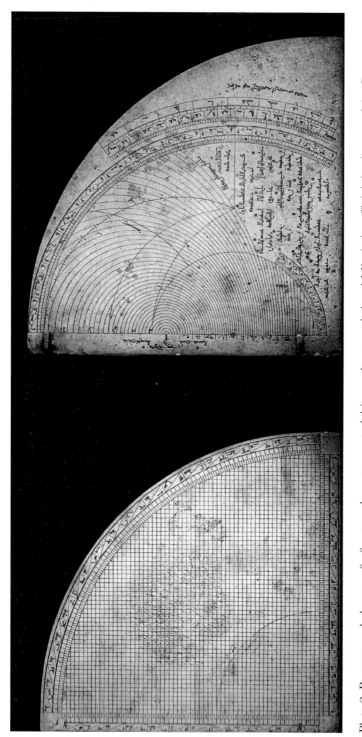

Plate 2 Reverse and obverse of a fourteenth-century astrolabic quadrant made by al-Mizzi, the official timekeeper of the Great Mosque of Damascus. British Museum.

Plate 3 Brass celestial globe, possibly from Maragha, Iran, made by Muhammad ibn Hilal in 674/1275–6. British Museum.

Plate 4 Brass astrolabe, perhaps Isfahan, late seventeenth–early eighteenth century. The Sarikhani Collection.

Plate 5 Steel mirror inlaid in gold and silver with the signs of the zodiac and their associated planets. Mamluk Syria or Egypt, c.1320–40. Treasury of the Topkapı Saray Palace, Istanbul.

Plate 6 Astronomers line up parts of an armillary sphere with specific stars to produce flat charts of the heavens for use in astrolabes. Sixteenth-century Ottoman manuscript, University Library, Istanbul.

Plate 7 Reconstruction by V. A. Nil'sen of the observatory of Ulugh Beg at Samarqand, 1428–9. Agence Rapho, Paris. © Roland & Sabrina Michaud/ Gamma-Rapho.

Plate 8 Astronomers at the observatory of Nasir al-Din Tusi at Maragha, NW Iran, being taught by means of the astrolabe. Manuscript of astrological treatises dated 813–14/1411, Shiraz, University Library, Istanbul.

Plate 9 The House of Saturn from a manuscript on astrology, for use in casting horoscopes. Iraq, probably Baghdad: late fourteenth or early fifteenth century (probably Jalayrid). The Keir Collection of Islamic Art on loan to the Dallas Museum of Art (collection owned by Pergamon Museum).

trained as a metalsmith. He entered the service of Qadi ibn Sa'id, ruler of Toledo, as a maker of astronomical instruments. His curiosity and intelligence, however, impressed his superiors, and he was soon supplied with treatises on mathematics and astronomy. Two years later (1062), having mastered the field, he became a member of the Toledan school of astronomers and soon after was appointed its head. He remained in the city for over twenty years – constructing astronomical instruments, recording observations of the Sun, Moon, planets, and stars, and composing treatises on astronomy and astrology. Forced to leave Toledo sometime before 1085 (when Alfonso VI of Castile finally conquered the city), he died in Cordoba in c. 1100.

While it was not uncommon for a medieval astronomer to construct his own instruments, al-Zarqali's achievements were of a different order altogether. He designed and built the famous water clock (see Glossary) of Toledo. Constructed of two basins and a series of connecting pipes, the clock generated a precise thirty-day lunar calendar. It worked perfectly until 1135 when Alfonso VII asked one of his astronomers to inspect it. Easily disassembled, its design and workmanship were so intricate that none of the ruler's men was able to rebuild it.

As an instrument maker Zarqali's fame was widespread, both in Andalusia and in the Latin West. He composed two works on the equatorium (see Glossary) – an instrument that allowed the astronomer to determine the positions of the Sun, Moon, and planets without resorting to tables or calculation. In the thirteenth century these treatises were translated into Spanish by order of Alfonso X and included in his *Libros del saber de astronomia (Books of the knowledge of astronomy)*. But his reputation as a craftsman was based primarily on his creation of the universal astrolabe – the Tablet of Zarqali (or al-Safiha al-Zarqaliyya). As the single most important instrument of the Muslim astronomer, the astrolabe had many uses. The planispheric astrolabe of the eleventh century, however, was severely limited: It could be used for a single latitude only. Al-Zarqali's creation, on the other hand, included a table of twenty-nine stars and could solve astrological/astronomical problems for any latitude. In medieval Europe, the Saphea Arzachelis (Disc of Zarqali) became widely popular.[8]

Despite his skill as an instrument maker, Zarqali's reputation, both in Andalusia and in medieval Europe, was based on his authorship of the *Toledan Tables*. Although this treatise included a good deal of material from earlier works, there is considerable evidence of fresh observations – both by al-Zarqali and others. Qadi ibn Sa'id had recruited a large group of Muslim and Jewish astronomers, and they conducted observations of the Sun, Moon, and planets for many years. Al-Zarqali joined them and was said to have observed the Sun for twenty-five years and the Moon for thirty. Although there was no mention in the sources of instruments or of an observatory, it is probably fair, given the number of astronomers and the extent of the work, to assume the existence of some kind of institution.[9]

Of the *Toledan Tables*, compiled in Arabic by al-Zarqali in c. 1080, no copy remains. From the Latin translation, done by Gerard of Cremona and entitled *Canones Azarchelis or Tabulas Toletanas*, it is evident that Zarqali's treatise was modelled on Khwarizmi's *Zij al-Sindhind*, with additional material from al-Battani, Ptolemy, and the author himself. Zargali's epoch was the Hijra and his meridian was that of Toledo. The solar, lunar, and maximum planetary equations were all from al-Battani. The planetary latitudes were from the *Zij al-Sindhind*, and the table of sines was from Brahmagupta's *siddhanta*. The astronomer reproduced several tables directly from the *Almagest* and updated the star catalogue he had inherited. By the early twelfth century both the *Toledan Tables* and the *Zij al-Sindhind* (in al-Majriti's adaptation) had become widely popular in medieval Europe. Although al-Zarqali's treatise was more accurate than al-Khwarizmi's, both were used by astronomers in England, France, and Italy – adapted for different latitudes and calendars. By the early thirteenth century, the *Toledan Tables* had begun to supersede the *Zij al-Sindhind*, holding sway until the advent of the Parisian *Alfonsine Tables* in the early fourteenth century.[10]

Al-Zarqali completed an almanac in 1089. Combining the Babylonian planetary cycles called goal years with Ptolemaic planetary values and parameters, the almanac provided detailed information on planetary positions without the need of additional computation. It included the true daily positions of the Sun for four years (1088–92) and the true positions of the five planets at five or ten day intervals for various periods – eight years for Venus, seventy-nine years for Mars, and so on. The trigonometric portion of the almanac exhibited the same mixture of sources and contained tables of sines, cosines, versed sines, secants, and tangents. Perpetual almanacs of this kind were a characteristic of Andalusian astronomy – common among astronomers and navigators for several centuries. Zarqali's almanac was translated into Latin in the mid-twelfth century and was widely influential, Regiomontanus perhaps the most important exponent of the Spaniard's approach.[11]

Zarqali's treatise on the Sun (entitled either *On the Solar Year* or *A Comprehensive Epistle on the Sun*) was completed in c. 1080. He had conducted systematic observations of the Sun for twenty-five years and was the first astronomer to establish a numeric value for the movement of the solar apogee. Previous astronomers, both Islamic and Greek, had thought that the solar apogee was a fixed value but Zarqali found that the apogee changed over time. He also determined its rate of change, calculating that it moved against the background of fixed stars and in the same direction as the zodiacal signs at the rate of 1 degree every 299 Julian years or about 12.04 seconds per year – an estimate that was quite close to the modern value of 11.6 seconds. He established a more accurate estimate for the solar eccentricity and confirmed the estimate in the *Toledan Tables* for the length of the sidereal year. His model for the motion of the Sun – the centre of the Sun's deferent moved in a small, slowly-rotating circle reproducing the movement of the solar apogee – was discussed by Regiomontanus and Puerback in the fifteenth century and employed by Copernicus in the sixteenth century in a heliocentric form.[12]

Zarqali's *Treatise on the Movement of the Fixed Stars* was composed in about 1085. In it he advanced a new theory of trepidation, namely, the theory that ascribed an oscillation to the slow westward movement of the background of fixed stars (or precession of the equinoxes). The reason for this apparent movement was the slow oscillation of the Earth's axis due to the gravitational force of the Sun and the Moon. The movement was very slow – the modern figure is one revolution every 25,800 years. But in early Islamic astronomy the figure was thought to be 49,000 years, and trepidation or oscillation in this movement was thought to occur every 7,000 years. Zarqali's model matched more closely the results of observation, and he was able to distinguish the variable precession of the equinoxes from the oscillation in the obliquity of the equinox.[13]

A short treatise by Zarqali (c. 1081) on the motion of the seven planets has become somewhat controversial. In Alfonso X's Spanish translation of the work – in a section entitled 'On the Invalidity of Ptolemy's Method to Obtain the Apogee of Mercury' – there was a diagram depicting a non-circular orbit for the planet Mercury. On this basis it has been argued that Zarqali antici-pated Kepler in stating that the planetary orbits were elliptical. Since at this point there is no evidence that Kepler ever saw Zarqali's treatise, the issue has remained unresolved.[14]

Zarqali devoted more than thirty years to a series of careful and systematic observations of the Moon, and he composed a treatise offering several cor-rections to Ptolemy's lunar model. This resulted in more accurate predictions of the appearance of the new Moon and of eclipses. These corrections were included in the Spanish version of the *Alfonsine Tables*. Zarqali also wrote an astrological work – *On the Motions and Influences of the Planets*. In the tradition of pre-Islamic Andalusia, this treatise described the use of magic squares to construct talismans.[15]

In the scholarship on Islamic astronomy most of the attention has been directed to developments in the East – in Syria, Iraq, and Iran under the early Abbasids and Seljuqs. In the West, under the Umayyads of Andalusia, the achievements were much less impressive, and their significance was due primar-ily to the fact that Spanish Islam was the principal conduit through which knowl-edge about astronomy – Greek, Indic, and Islamic – reached the Latin West. Thus, having finished our account of the two principal Andalusian astronomers – al-Majriti and al-Zarqali – it is perhaps time to turn to a comparison of Islamic astronomy (East and West) in the first century of the second millennium.

Al-Zarqali and 'Umar Khayyam were contemporaries, doing most of their work in the last half of the eleventh century. However, in Isfahan 'Umar and al-Khazini enjoyed superior facilities and had greater support than did al-Zarqali and his fellow astronomers in Toledo. In Isfahan, Malik Shah constructed a new observatory, collected a large group of immovable instruments, and set aside an on-going source of funds. In Toledo, on the other hand, Qadi ibn Sa'id, ruler of a much smaller principality, had fewer resources at his disposal, and, as a result, his astronomers did not have access to the newer instruments or to long-term financial support. And these differences in organisation and money were reflected in their respective publications, the astronomical treatises produced in the two cities differing greatly in breadth and originality.

Although al-Zarqali observed the Sun for twenty-five years and the Moon for thirty, the *Toledan Tables* were mostly a pastiche of borrowings. Based on the *Zij al-Sindhind*, Zarqali's treatise incorporated tables from al-Battani, al-Khwarizmi, Brahmagupta, and Ptolemy. Very little was the result of fresh observation and original calculation. On the other hand, the *Zij al-Mu'tabar al-Sanjari* of al-Khazini (and to a lesser extent the *Zij-i Malik Shahi* of 'Umar Khayyam) contained tables of original observations by more astronomers working over a longer period with more precise instruments. And, as a result, the Isfahan treatise was much more accurate and comprehensive.

The chronological section in the *Toledan Tables* was basic – the Hijra era and epoch replaced the Yazdegird. For 'Umar Khayyam and al-Khazini, on the other hand, chronology was a central concern. 'Umar had deflected Malik Shah's desire for a completely new *zij* by promising him a new era named after himself. Not only was the *Tarikh-i Jalali* the result of original observation and careful recalculation – returning Naw Ruz to the vernal equinox and introducing a new epoch – but it also produced a new solar era and calendar. An innovation that was indispensable for an expanding agrarian empire, the *Tarikh-i Jalali* solved economic and administrative problems as well as astronomical. Al-Khazini also devoted a considerable space to chronology in his treatise. In addition to the Hijra era, he included descriptions of the Yazdegird, Jewish, and Indic eras and calendars along with instructions on translating dates from one era and calendar to another. He included the dates of Muslim, Christian, Jewish, and Zoroastrian feasts and festivals, regnal tables for dynasties past and present, and tables listing the names and dates of the major prophets.

In al-Zarqali's *Toledan Tables* the solar, lunar, and planetary tables were all taken without revision from al-Battani's early tenth-century *Zij al-Sabi*. In Khazini's *zij*, on the other hand, the tables for lunar mansions were new and those for the mean motions of the Moon and the five planets were fresh and precise (worked out to seven places) – the result of a long-term programme of observation by the resident astronomers (working with new instruments) at Malik Shah's Isfahan Observatory. The planetary tables included equations, stations, latitudes, and retrogrades. Al-Zarqali devoted more than thirty years to observing the Moon and was able to correct several of Ptolemy's lunar figures.

Al-Khazini, on the other hand, produced completely new lunar visibility tables – the most extensive yet, replacing entirely those of Ptolemy and al-Battani. While al-Zarqali updated as best he could Ptolemy's star tables, al-Khazini conducted fresh observations and produced new coordinates and brightness estimates of forty-six stars in 1106–7.

Because of its connection to astrology, Islamic astronomy was almost exclusively concerned with planetary motion – with plotting and predicting the positions of the Sun, Moon, and the five planets. A surprising result of this pre-occupation was an extremely narrow focus. For example, the Andalusian and Seljuq astronomers completely ignored the supernova of 4 July 1054. The result of an exploding star, this supernova was four times brighter than Venus and was visible in daylight for twenty-three days and at night for nearly two years (653 days). The Chinese astronomers took careful note of the event, recording its appearance and describing its evolution in great detail. Yet al-Zarqali (26 years old) and the Toledan astronomers did not mention it; nor was is referred to in any of the works of 'Umar Khayyam and the astronomers of the Isfahan Observatory.[16]

Although the *Toledan Tables* of al-Zarqali and his colleagues were not as comprehensive or as accurate as the treatises of 'Umar Khayyam and al-Khazini, the Andalusians were not without an impact on Islamic astronomy. Zarqali's contributions to the syncretic tradition of Andalusia (part Sindhind, part al-Battani, and part Ptolemy) were significant. A new theory of trepidation, a determination of the motion of the solar apogee, a new solar model with variable eccentricity, and multiple corrections to Ptolemy's lunar model – all of these were influential in both the Islamic East and the Latin West.[17] Copernicus had studied al-Zarqali and the Toledan school, embracing the Andalusian's theory of trepidation, his variable length of the tropical year, and his changing value for the obliquity of the ecliptic. The sixteenth-century astronomer, however, had a different objective in mind. Whereas Zarqali and his colleagues were interested in calculating the positions of the heavenly bodies in order to draw up horoscopes, Copernicus was concerned with constructing a model of the real world, creating an astronomical system that reflected physical reality.[18] Al-Zarqali's stature in early European astronomy is reflected in the name given by Giovanni Riccioli (in 1651) to one of the most prominent of the Moon's craters – Arzachel.

The last creation of Islamic astronomy in Andalusia was the work of Alfonso X (1252–84), Christian ruler of Castile, Leon, and Galicia. A man of many interests, Alfonso introduced the first vernacular law code in Spain and created the Mesta, an association of sheep farmers in the central plain. In the imperial election of 1256 he mounted a brief but unsuccessful attempt to succeed William II of Holland as Holy Roman Emperor. The most important aspect of his reign, however, was his campaign to establish Castilian (the ancestor of modern Spanish) as the language of science and literature in Christian Spain, replacing Latin.

As a patron of literature and learning, Alfonso expended a great deal of time and money in an effort to recover eastern Islamic and Andalusian astronomical and astrological materials by translating them from Arabic into Castilian. For this purpose he founded the *Toledan School of Translators* which included one Muslim convert, eight Christians (four Spaniards and four Italians), and five Jews (in particular Yehudah ben Mosheh and Isaac ibn Sid called Rabicag, authors of the *Alfonsine Tables*) but no Muslims. Due to the animosity created by the Reconquista, Alfonso was not able to convince a single Andalusian astronomer to join his group. His plan for the school was ambitious: to create two large collections of astronomical/astrological works, both translations and original compositions. The first collection focused on magic and included the *Picatrix* (a handbook on talismanic magic), the *Lapidario* (a book on the magical applications of stones), and the *Libro de la Magica de los Signos (Book on the Magic of Signs)*.

The second and much greater collection was given over to works on astronomy and astrology. It included the well-known *Libros del Saber del Astronomia (Books of Astronomical Knowledge)*; Ibn al-Haytham's *Configuration of the Universe*; the canons or instructions for al-Battan's *Zij al-Sabi*; the *Alfonsine Tables*; a treatise on the use of the sine quadrant; Ptolemy's *Tetrabiblos*; al-Rijal's treatise on astrology (*The Complete Book on the Laws of the Stars*); and the anonymous *Libro de las Cruzes (Book of the Crosses)*.[19]

The first book in the *Libros del Saber del Astronomia* was a treatise on celestial cartography based in part on al-Sufi's *Book of Fixed Stars*. The other works in the *Libros* were devoted to astronomical instruments. One group contained treatises on analogical calculators – the celestial sphere (see Glossary), the spherical and plane astrolabe, and the *Saphea Arzachelis* or *Disc of Zarqali*. The main purpose of these instruments was to provide graphic solutions to problems associated with horoscopes. A second group was devoted to timekeeping and contained works on the sundial and water clock. There were also two works on the equatorium. For each of these instruments Alfonso required two compositions: one on construction and the other on usage. If an adequate Arabic treatise was available, it was translated and included. Otherwise, an original work was drafted, usually by Rabicag. In addition to the *Libros*, there was an assortment of volumes given over to astrology: the *Book of the Crosses* and the treatises by Ptolemy and al-Rijal. These texts offered the astrologer principles for interpreting horoscopes – for predicting the future of individuals and dynasties and for determining auspicious or inauspicious times for battles, journeys, and marriages.

Of all the astronomical treatises translated or created by the astronomers of Alfonso's school, the most important by far was the *Alfonsine Tables*. Alfonso's treatise, however, has generated a great deal of controversy. Two different versions appeared – one in Castilian (c. 1271) and the other in Latin (c. 1327)

– and the controversy concerned the relationship between them. The Castilian version, written by Yehudah ben Mosheh and Rabicag, contained only the canons, the written materials that explained how to use the numerical tables. The tables themselves were missing.[20] In their introduction to the canons Yehudah and Rabicag wrote:

> Two hundred years have passed since the time of Al Zaraqali's observations, and in his tables certain divergences have appeared which are discernible and well-known to those who have clear vision, so that there can be no excuses for overlooking them.
>
> At this junction came the felicitous and propitious reign and the monarchy of the very high and noble Don Alfonso, may God keep him. As he honored and protected people of learning, he caused the preparation of the instruments mentioned by Ptolemy in his Almagest of the type of armillas and other instruments and ordered us to make astronomical observations in the city of Toledo . . . in which the observations of Al Zaraqali had taken place.
>
> He called upon us to rectify the divergences which were noticeable in the positions of certain planets and in other movements. We complied with his order as it was fit. We improved (prepared) the instruments so as to make them as perfect as possible and busied ourselves with observations for one season, extending those of the Sun to a full year. We further observed the Sun at equinoxes and solstices, starting our observations before the Sun's arrival at those points and continuing them for some time after, as well as in parts of the heavens at the centres of Taurus, Scorpio, Leo, and Aquarius.
>
> We also made observations of a few conjunctions of planets with other planets and with fixed stars. We, likewise, observed many solar and lunar eclipses, and we had recourse to still other observations whenever we were in doubt and repeated them many times in order to remove the uncertainties.
>
> We left nothing unexplored, and we continued our scrutiny until all that needed improvement was improved. And as everything was checked, we have accepted as correct everything that was certain or nearly certain and have constructed these tables on the basis of results obtained from these observations.[21]

In Alfonso's day all models of the solar system were Ptolemaic, and, as a result, Yehudah and Rabicag would have employed a geocentric theory of planetary motion based on a complex system of epicycles. A famous but apocryphal story has Alfonso, after hearing a complicated mathematical explanation of Ptolemy's model, remarking 'If the Lord Almighty had consulted me before embarking on creation, I should have recommended something simpler.'[22]

Because the tables of Alfonso's Castilian version are missing it was difficult to decide how much of his programme was actually carried out. Of actual observations, the only evidence was of three lunar eclipses (one in 1265 and two in 1266) and one solar eclipse (1263). The other numerical parameters in the canons or in various Alfonsine materials were all from the *Toledan Tables* of al-Zarqali. However Jos Chabas and Bernard Goldstein, authors of a recent book on the

Castilian version of the *Alfonsine Tables*, have put forward some persuasive arguments about the probable contents of the lost tables.[23] And in a long book review N. M. Swerdlow has summarised their position, supporting and supplementing their conclusions.[24]

A comparison of two early versions of the Latin *Alfonsine Tables* convinced Chabas and Goldstein (as well as Swerdlow) that the quantitative materials in the early-fourteenth century tables probably originated in the late thirteenth century Castilian version. With this hypothesis in hand, it is possible to suggest an outline of the contents of the tables in the original treatise of Yehudah and Rabicag. The equation tables were at intervals of one degree for the Sun, Moon, and planets and were descendants of Ptolemy's *Handy Tables*, transmitted by way of al-Battani and the *Toledan Tables*. There were, however, alterations in the maximum equations of anomaly for the Moon, Sun, Venus, and Jupiter. The tables for latitude followed the form of the *Almagest* rather than that of the *Handy Tables*, although at six degree intervals as in al-Battani and the *Toledan Tables*. There were tables for eclipses (the mean and true conjunction and opposition), tables for solar and lunar velocity, shadow tables, and tables for spherical astronomy (equation of time, declinations, and houses). A star catalogue based on Ptolemy's was included with some Arabic star names. The locations of the stars were precessed by 17 degrees, 8 minutes and the era was Alfonsine – epoch 1 June 1252. The tropical year was constant and the sidereal year variable.

The most distinctive feature of the *Alfonsine Tables*, however, was the method of computing mean motions for eras and calendars. The mean motion tables themselves were purely sexagesimal, giving the motions for one to sixty days from six to eight places. Ptolemy's tables in the *Almagest* were to six fractional places although with decimal integers, and he extended his tables for daily motion from thirty to sixty days. Accompanying these were tables for the epochs of the different eras and the mean positions were for the meridian of Toledo. The problem, however, was the discrepancy between the lost tables described in the Castilian canons and the actual tables encountered in the early fourteenth century Latin treatises. The mean motion tables implied in the Castilian canons were for the meridian of Toledo in decimal not sexagesimal form and for the Julian era and calendar alone.

Despite these differences Chabas and Goldstein argued (and Swerdlow agreed) that the quantitative tables in the Latin *Alfonsine Tables* likely derived from the earlier Castilian treatises of Alfonso. In Toledo in the thirteenth century there were in all probability many versions of the *Alfonsine Tables*.[25] It is only an accident of history that the lone surviving manuscript employed decimal tables of mean motion rather than sexagesimal. As summarised by Swerdlow, there were strong arguments for this position. In the first place, the early Latin tables were all called *Alfonsine*. Why would they have adopted such a name unless there was an actual connection. Secondly, the sexagesimal tables in the earliest Parisian tables were for the meridian of Toledo not Paris. Again, why? The sexagesimal tables of the Latin version were for a variety of eras – the Spanish, the Incarnation, the Hijra, the Yazdegird, and the Alfonsine – and for a variety of calendars – the Julian, Islamic, and Zoroastrian. And there were rules for translating dates from one calendar and era into another. In Toledo such a variety would have been relevant but in Paris what possible interest could there have been in the Yazdegird era and the Zoroastrian calendar?

The European version of the *Alfonsine Tables*, with canons in Latin and a meridian of Paris, was first compiled by a group of French astronomers – John of Saxony, John of Murs, and John of Ligneres. These tables were the most influential astronomical work of the later Middle Ages. For more than two hundred years, adapted for different meridians under various names and with various modifications and additions, they formed the basis of all astronomical treatises in late Medieval and early Renaissance Europe. They were the standard until Erasmus Reinhold published his *Prutenic Tables* in 1551. Nicholas Copernicus used parameters from the *Alfonsine Tables* in his early *Commentariolus* (*Little Commentary*) and the Alfonsine year of 365 days, 5 hours, 49 minutes, and 16 seconds was the mean tropical year for his *On the Revolutions of the Heavenly Spheres*. It also became the basis for the Gregorian Reform of the Julian Calendar (1582).

It is important, however, to realise that in the history of Islamic astronomy the *Alfonsine Tables* represented an early stage. They were done by two astronomers working for a relatively short period (1263–72) with small, mostly portable instruments. The equipment, facilities, and approach were more like those of al-Ma'mun's astronomers at Shammasiyya and Mt Qasiyun than like those of 'Umar Khayyam at the Isfahan Observatory of Malik Shah. The *zij* produced by Alfonso's two astronomers was mostly derivative – modelled after al-Majriti's verson of al-Khwarizmi's *Zij al-Sindhind* with parameters from al-Battani, Ptolemy, and Brahmagupta. Fresh observations and new calculations were limited. And, as we shall see in the next chapters, the glories of Islamic astronomy were still ahead. Although the Maragha Observatory (1259–62) of Nasir al-Din Tusi and his *Zij-i Ilkhani* (1272) were contemporaneous with the Toledo School and the *Alfonsine Tables*, the Samarqand Observatory (1420) of Ulugh Beg and his *Zij al-Sultani* (1440) and the Istanbul Observatory (1577) of Taqi al-Din and his *Zij-i Shahinshahi* (c. 1578) were well in the future. The great institutions of the Eastern Islamic world – in Maragha, Samarqand, and Istanbul – boasted dedicated observatories, staffed with mathematicians and astronomers working on larger and more precise instruments, and producing more accurate tables employing more creative theories. Because of the Reconquista, however, from the early thirteenth century onward the link between Andalusia and the Eastern Islamic world was broken (witness Alfonso's inability to recruit even a single Islamic astronomer). As a result, knowledge of the crowning achievements of

Muslim astronomers and mathematicians was unavailable to the Christian courts of Andalusia after the reign of Alfonso X. And since Andalusia was the principal source of European knowledge of the Muslim world, the Islamic astronomy which reached the Latin West and which held sway until the early sixteenth century was mostly the outdated work of the Toledan astronomers of the eleventh-thirteen centuries – al-Zarqali and the Jewish Alfonsine authors. Thus, until Tycho Brahe, Kepler, Galileo, and Copernicus in the sixteenth and seventeenth centuries, European astronomy remained well behind that of the Islamic world – in organisation, equipment, and scientific creativity. The Latin *Alfonsine Tables* of the early fourteenth century held sway for the next two hundred years, and the achievements of Tusi, Ulugh Beg, and Taqi al-Din remained largely unknown to the fledging astronomers of early Renaissance Europe.

Notes

1. Lapidus, *History of Islamic Societies*, 378–89.
2. Julio Samso, *Islamic Astronomy and Medieval Spain* (Aldershot: Ashgate Publishing Ltd, 1994), 86–7.
3. Ibid. 235–43.
4. Ibid. 80.
5. Ibid. 229–30.
6. Ibid. 5.
7. 'al-Majriti, Abu al-Qasim Maslama Ibn Ahmad al-Faradi', *Dictionary of Scientific Biography*.
8. North, *Cosmos*, 218–21; 'al-Zarqali, Abu Ishaq Ibrahim', *Dictionary of Scientific Biography*.
9. Sayili, *Observatory*, 181.
10. North, *Cosmos*, 221–3; Kennedy, *Islamic Astronomical Tables*, 128–9.
11. Saliba, *A History of Arabic Astronomy*, 20.
12. Ibid. 16–17.
13. Ibid. 14–16; 'al-Zarqali, Abu Ishaq Ibrahim', *Dictionary of Scientific Biography*.
14. 'al-Zarqali, Abu Ishaq Ibrahim', *Dictionary of Scientific Biography*.
15. Ibid.
16. M. Heydari-Malayeri, 'The Persian-Toledan Astronomical Connection and the European Renaissance', Academia Europaea 19th Annual Conference: The Dialogue of Three Cultures and our European Heritage (Toledo Crucible of the Culture and the Dawn of the Renaissance), 2–5 September 2007, Toledo, Spain.
17. Saliba, *A History of Arabic Astronomy*, 22–3.
18. Ibid. 18–19.
19. Hockey, *Biographical Encyclopedia of Astronomers*, 29–31.
20. Ibid.
21. Sayili, *Observatory*, 365–6.
22. Emmanuel Poulle, 'The Alfonsine Tables and Alfonso X of Castile', *Journal of the History of Astronomy* 6 (1988): 97.
23. Jos Chabas & Bernard R. Goldstein, *The Alfonsine Tables of Toledo* (Boston and London: Kluwer Academic Publishers, 2003).
24. N. M. Swerdlow, 'Alfonsine Tables of Toledo and Later Alfonsine Tables', *Journal of the History of Astronomy* 25 (2004): 479–84.
25. For a discussion see Saliba, *A History of Arabic Astronomy*, 33ff.

CHAPTER FIVE

The observatory in Maragha

In the history of Islamic astronomy the thirteenth century was the most important. It witnessed the founding of the Maragha Observatory, the most advanced
scientific institution in the Eurasian world. Lavishly funded, Maragha had
an extensive library and a large staff. And its head astronomer, Nasir al-Din
Tusi (1201–74), compiled the most complete and up-to-date *zij* yet available,
while also composing ground-breaking works in mathematics and astronomy.
Because, however, the Reconquista had ended intellectual exchange between
the Islamic and Christian worlds, the discoveries of Tusi and the Maragha
astronomers remained unknown in Europe for more than two hundred years.

Nasir al-Din Tusi was one of the great polymaths of medieval Islam. He
wrote more than one hundred and fifty works in both Persian and Arabic,
covering an impressive range of subjects. Expert in both the traditional and the
rational sciences, he completed treatises on law, Shi'ite theology, Sufism, logic,
ethics, medicine, and metallurgy. He wrote critical summaries of the major
Greek mathematicians and philosophers and composed original works in arithmetic, geometry, trigonometry, astrology, and astronomy. His fame stretched
from Baghdad in the West to China in the East. Known in his time as 'Khwaja'
(distinguished teacher), he was later given the title of 'third teacher' – after
Aristotle and Abu Nasr Muhammad ibn Muhammad al-Farabi (872–950). Ibn
Khaldun considered him the greatest of the later Persian scholars.[1]

Born into a family of Twelver, or Imami, Shi'ites in Tus, a town in eastern
Iran, Nasir al-Din studied the traditional sciences with his father, a scholar of
Imami law, logic, and natural philosophy, and metaphysics with his uncle. In
Tus he also received basic instruction in algebra and geometry. In 1213 he
travelled to Nishapur to read philosophy with Farid al-Din al-Damad and medi-
cine with Qutb al-Din al-Masri. Later during the 1220s he travelled to Mosul
in northern Iraq to study mathematics and astronomy with Kamal al-Din ibn
Yunus (1156–1242).

By the time he had finished his formal education the Islamic world had been
thrown into chaos by the Mongol hordes of Chinghiz Khan (1206–7). The
steppe warriors of the great Khan had conquered an enormous swath of territory from the Pacific in the East to the Caspian in the West. After Chinghiz's
death his sons continued to advance, adding Russia, Iran, and south China
to what became the greatest land empire ever seen. In eastern Iran the only

islands of peace were the mountain fortresses of the Isma'ilis. A Shi'ite sect, the Isma'ilis, or Seveners, differed from the Imamis, or Twelvers, in their allegiance to the seventh Shi'ite Imam as the true descendant of 'Ali. The largest and most successful Isma'ili state had been the Fatimid caliphate (c. 909–1171), which ruled Egypt, North Africa and parts of Syria and the Arabian peninsula from the tenth through the twelfth centuries. In Iran, a smaller Isma'ili state was centred on the mountain fortresses of Alamut and Lambsar. Founded by the Isma'ili leader Hasan al-Sabah in 1090, Alamut was virtually impregnable.

Sometime after 1230, the Isma'ili ruler, Nasir al-Din Muhtashim, invited Tusi to relocate to Alamut. Over the next twenty-five years he wrote many of his most famous works – in philosophy, logic, and astronomy – achieving a sizeable reputation. In 1256 the Mongol general Hulagu (1256–65), grandson of Chinghiz Khan and founder of the Ilkhanate state in Iran, defeated the Isma'ili forces and seized Alamut. Because of Tusi's fame and Hulagu's interest in astrology, the Iranian quickly became a valued member of the Ilkhanid court. Appointed chief astronomer and superintendent of charitable endowments (*awqaf*, sing. *waqf*), he accompanied the Mongol army in its victorious march west, witnessing Hulagu's capture of Baghdad in 1258. In 1259 the Ilkhanate ruler decided to underwrite the construction of a large observatory in his new capital of Maragha, a city in the north-western Iranian province of Azerbaijan. Appointed its first director, Tusi oversaw the installation of new instruments, the stocking of a large library, the recruitment of a talented staff, and the compilation of a new astronomical treatise – the *Zij-i Ilkhani*. In 1274 Nasir al-Din and a group of students journeyed to Baghdad. Soon after his arrival he fell seriously ill and several weeks later he died. He was buried a few miles outside the city walls, near the mausoleum of Musa al-Kazim, the seventh Twelver Imam.

Tusi's works spanned an incredible range. His first major composition, finished during the early 1230s when he was barely twenty, has remained his most popular. *Akhlaq-i Nasiri* (*The Nasirian Ethics*), dedicated to the Isma'ili ruler who first offered him sanctuary, was a work of theory. It was a subtle blending of the philosophical and scientific theories of Aristotle and Plato, on the one hand, with the Islamic view of man, society, and the universe, on the other. It was divided into three parts. Part One, on ethics, analysed principles and ends: the nature of man, his virtues and vices. Part Two, on economics, dealt with the organisation of the household: the regulation of wives and children, the rights of parents, and the government of slaves and servants. Part Three, on politics, examined the structure of society: the divisions between classes and the government of cities and states. For the Muslims of India and Iran *Akhlaq-i Nasiri* has remained the most popular work on ethics.[2]

Over the next several years Tusi's interests turned more toward the rational sciences. In 1235 he wrote his first astronomical work, *Risala-yi Mu'iniyya* (*Treatise for Mu'in [al-Din]*). An elementary effort intended for students, it was purely descriptive and did not take up the problems of Greek astronomy. During the following decade, however, he delved more deeply into mathematics and astronomy. In 1247 he produced his recension of Ptolemy's *Almagest* – *Tahrir al-Majisti*. In it he sketched a first attempt at solving a major problem in Ptolemy's model of planetary motion. In order to make his model match his observations, Ptolemy had created a complicated system of epicycles, a nesting of circles within circles that attempted to track and predict the movements of the planets. To accomplish this he was forced to introduce a second epicyclic centre, the equant, which was near but not identical with the Earth's centre. Since Ptolemy's various epicycles did not share a common centre, it was obvious to Arab astronomers that his model was fatally flawed: it violated the principle of uniform circular motion.

Tusi's solution was a mathematical theorem – the so-called 'Tusi Couple'. Given a preliminary exposition in the *Tahrir al-Majisti*, the theorem was fully proved in two later compositions – the 1255 *Solution of the Difficulties in the Mu'iniyya*, an addendum to the earlier *Treatise for Mu'in al-Din* and the 1261 *Treatise on the Science of Astronomy*. In order to retain Ptolemy's accuracy, while at the same time eliminating his violation of the principle of uniform circular motion, Tusi proposed a modification. His new model contained two circles or spheres, one twice the size of the other. Both revolved around an identical centre, the smaller moving at twice the speed of the larger. Tusi showed that the movement of the two spheres would cause the common centre to oscillate along the diameter of the larger sphere, thus producing the required linear motion as the result of two uniform circular motions.[3]

Like all Islamic astronomers, Tusi was a practicing astrologer. He composed an elementary work in thirty chapters on the topic of determining the positions of the planets. Its full title was *A Short Treatise on Constructing an Almanac* but it was popularly known but the shortened version, *Thirty Chapters*.

Although Tusi composed original compositions in both astronomy and mathematics, some of his most valuable scientific work was found in his critical editions of the Greek classics. A '*tahrir*', a redaction or recension, involved much more that simply comparing and collating the various manuscripts of a given work. It meant commenting and correcting, eliminating repetitions, providing clarifications, and offering substitutions. Tusi's work in this area was remarkable. He produced the standard redactions of the core works of the scientific curriculum. Beginning with Euclid's *Elements* (geometry), he continued through the works of the principal Greek mathematicians (Autolycus, Aristarchus, Apollonius, Archimedes, Hesicles, Theodosius, and Menelaus), ending with Ptolemy.

The most important of his redactions, and the one that best illustrates his method, was *Tahrir al-Majisti* (*Redaction of the Almagest*), completed in 1247 during his sojourn in Alamut. Tusi began by comparing the various Arabic translations

of Ptolemy's famous work, substituting a word or phrase when necessary, eliminating repetitions, and replacing the older terminology with newer expressions.
Where the text was obscure or faulty, he would point out the problems and
either correct the errors immediately or promise a solution in later publications.

In logic, Tusi composed five works. His most important, *Foundations of Inference*,
also written in Alamut, was the most extensive treatment of the subject since Ibn
Sina. Although Tusi respected his predecessor's achievement, he surpassed him
in analysing the relationship between logic and mathematics. He converted
logical terms into mathematical signs and distinguished between the meanings
of substance in both the philosophical and the scientific senses. He also clarified
the relationship between certain categories in metaphysics and logic.[4]

After moving to Maragha with Hulagu and beginning construction on the
new observatory, Tusi composed in 1261 his principal work on astronomy – the
Treatise on the Science of Astronomy.[5] Undertaken at the request of a colleague, it
contained the latest and most thoroughgoing critique of Ptolemy's planetary
model. In the treatise Tusi presented his final version of the 'Tusi Couple', the
only new mathematical theory to appear in medieval Eurasian astronomy, and
the one that influenced not only Qutb al-Din Shirazi and Ibn al-Shatir but also
Regiomontanus and Copernicus. Tusi also offered his opinion on the Milky
Way. Some 350 years before Galileo and the telescope he wrote:

> The Milky Way (the galaxy) is made up of a very large number of small tightly
> clustered stars, which, on account of their concentration and smallness, seem to
> be cloudy patches, because of this, it was likened to milk in colour.[6]

Immediately popular, the *Treatise* acquired a large readership and many
commentaries and soon became the standard work on astronomy in medieval
Islam. Like so many other astronomers, Tusi was interested in the tools of his
trade and composed a treatise on the construction and operation of the astrolabe, *Treatise on the Astrolabe*.

In addition to his work on astronomy, Tusi also composed original works
in arithmetic, geometry, and trigonometry. His *Computation with Board and Dust*,
written in Maragha (1264), dealt with computational mathematics. He analysed
ordinary arithmetical operations, combining traditional Greek number theory
with Indian numerals and techniques. He also introduced a new method for the
extraction of higher roots. Since, like most astronomers, he depended on the
patronage of the gullible and superstitious, he devoted a section of his treatise
to the arithmetic of astrology. He explained the arithmetic involved in casting
horoscopes: how to calculate signs and specify the locations of the heavenly
bodies using the sexagesimal numeral system.[7]

In geometry, Tusi followed the lead of 'Umar Khayyam. In his *Satisfying
Treatise*, Tusi analysed Euclid's fifth postulate (the parallel postulate). Although
he, like Khayyam, was unable to completely prove the postulate within the

confines of the Euclidean system, his work suggested some of the elements that would appear in the non-Euclidean geometries of the future.

Tusi has also been given credit for establishing trigonometry as an independent branch of pure mathematics. In his redaction of the *Almagest*, he replaced the chord theorems of Menelaus and Ptolemy with their trigonometric equivalents. And in *Book of the Complete Quadrilateral*, he gave an extensive description of spherical trigonometry – one that was independent of astronomy. He was the first to list all six cases for a spherical right triangle. He discovered the law of tangents for spherical triangles, and he stated the law of sines for both plane and spherical triangles. In his treatise Tusi presented an overview of trigonometry that was essentially modern, not much different from the account given in modern textbooks.[8]

Tusi compiled a treatise on mineralogy, *Book of Precious Metals*. Composed in Persian, it took al-Biruni's earlier work (*Book of Precious Stones*) as its model and occasional source. Tusi's work contained four chapters. The first treated the nature of compounds: the four elements, and their mixtures. The second was devoted to the qualities and properties of jewels. The third took up metals and the fourth perfumes.[9]

Of the sciences, medicine seemed to have held the least interest for Nasir al-Din. He composed a treatise on the most famous medical text of his day, Ibn Sina's *Canon of Medicine*, and he wrote his own work (*Principles of Medicine*). Tusi's interest in the field, however, was mostly philosophical, and his greatest contribution was in the area of psychosomatic ailments.[10]

In addition to *Akhlaq-i Nasiri*, Tusi wrote other works on philosophy. One of the most important was *A Commentary on the Isharat*, an interpretation of Ibn Sina's *Remarks and Admonitions*. Although Tusi was mainly interested in supporting Ibn Sina and rebutting the criticism of Fakhr al-Din Razi, he asked new questions in his commentary – about the nature of space and the creation of the physical world, among others. His philosophical works were ordinarily written in Persian, and he also left a collection of Persian poetry. As a result, he has been considered one of the pioneers in the development of modern Persian.[11]

Tusi was a member of the Imami (or Twelver) branch of Shi'ite Islam and in theology, as in everything else, his impact was considerable. His *Summation of Belief* was the first systematic treatment of Shi'ite *kalam* (speculative theology) and became the foundation for all later theological speculation.[12]

A great deal of Nasir al-Din's scientific work was done in the observatory built by the Ilkhanid ruler Hulagu Khan (1256–65). When Hulagu overran the Isma'ili capital of Alamut in 1256, Tusi was 52 and widely famed as an astronomer, mathematician, and philosopher. Although not much is known about his work as a practicing astronomer before 1259, when construction began at Maragha, one historian reported the presence of globes, armillary spheres, and astrolabes in the Isma'ili fortress at the time of Hulagu's conquest. However, none of the historical accounts mention organised observational activity, and there is no evidence that an astronomical treatise (*zij*) had been compiled. Nevertheless, it is tempting to regard the Alamut years as Tusi's introduction to the practice, as opposed to the theory, of astronomy – a first experience that whetted his appetite for the real thing.[13]

About Hulagu's decision to found an observatory in his new capital of Maragha, two stories were current. The first tied the initiative to Hulagu's brother, the great khan Mongke (1251–9), grandson of Chinghiz. Mongke had a passionate interest in mathematics and astronomy and he decided to found an observatory – either in Beijing or in Qaraqorum, the Mongol capital. Told that Nasir al-Din was the man to oversee the project, Mongke asked Hulagu to send the astronomer as soon as the Isma'ilis were defeated. Hulagu, however, greatly impressed by Tusi's personality and credentials (as an astrologer as well as an astronomer and mathematician), concluded that the observatory should be built in Iran rather than in China or Mongolia.[14]

The other story assigned the motivation to Tusi himself. Soon after being appointed chief astronomer at the Ilkhanate court, he is said to have persuaded Hulagu to build a new observatory. For Hulagu, as for many other rulers of the time, the attraction of such an institution was more astrological than astronomical. Hulagu was also interested in alchemy, and several anecdotes illustrate Tusi's skill at managing the ruler's superstition. Once, worried about the cost of the project, Hulagu questioned Tusi about the value of astrology. Since prediction is based on the unalterable character of the future, nothing can be done to alter what is predestined. Tusi replied with a story:

> Suppose you order someone to drop a large object from a very high place. This will produce a very loud noice and frighten all the people standing nearby . . . Only you and the person who carried out your order will remain perfectly calm, for only you both knew what would happen.[15]

At another point Hulagu ordered the execution of the famous historian 'Ata al-Mulk Juwaini (1226–83), author of *History of the World Conqueror*. Juwaini's brother asked Tusi to intervene. Although it was impossible for the Khan to countermand his order, the astronomer replied, all was not lost. The next day Tusi walked toward Hulagu's encampment with a staff and an astrolabe, followed by a man swinging a censer of incense. Tusi inquired about Hulagu's health and, told that it was satisfactory, fell to his knees in thanksgiving. After repeating the inquiries and acts of gratitude, the astronomer asked to see the Khan. He told the ruler that his horoscope was ominous, foretelling impending disaster. To ward off the calamity Tusi had burned incense and offered prayers but more was needed. The Khan should perform an act of great charity – free those in prison under sentence of death. The Khan agreed and Juwaini was saved.[16]

The construction of the buildings and the manufacture of the instruments were entrusted to Mu'ayyad al-Din 'Urdi (c. 1200–66). An engineer and astronomer, 'Urdi moved to Damascus sometime before 1239, teaching geometry and astronomy and designing astronomical instruments for the Mamluk amir, Malik al-Mansur. 'Urdi wrote two treatises on astronomy. In *Book of Astronomy* he, like Tusi, criticised Ptolemy's planetary models. Like Tusi he was bothered by the inconsistency between the theoretical mathematical models, on the one hand, and the physics of real physical bodies, on the other. 'Urdi's solution, the "Urdi Lemma', transformed eccentric models into epicyclic ones. Although his model differed from the 'Tusi Couple', it remained influential and was eventually adopted by Copernicus as a model for the movement of the upper planets. In the late 1250s Tusi invited 'Urdi to come to Maragha and supervise the construction of the observatory and the design and installation of the instruments. His second astronomical treatise, *A Treatise on Instrumentation*, was a rich and informative work describing the astronomical instruments at Maragha.[17]

The Maragha Observatory was located outside the city on the flattened top of a hill – 400 metres long and 150 wide. In order to lift water to the site 'Urdi designed a complex system of water wheels and acqueducts. He also oversaw the construction of a residence for Hulagu and a mosque. A small structure nearby contained a large dome. The rays of the Sun entered a hole in the dome, enabling the Maragha astronomers to measure both the elevation (at different times and seasons) and the mean motion (in degrees and seconds) of the Sun. The building was oriented so that the solar rays hit the threshold of the dome precisely on the vernal equinox (21 March). On its inner walls were illustrations: of the celestial spheres, the phases of the Moon, and the signs of the zodiac. Here also were terrestrial and celestial globes, maps of the seven climes, and graphs showing the length of the days and nights. The original terrestrial globe was made of paper pulp but 'Urdi's son fabricated a later version (1279–80) out of metal.[18]

The principal instruments, however, were located out of doors. 'Urdi began work in 1259, soon after the cornerstone was laid, and was mostly finished by 1262.[19] The first instrument was a mural quadrant (graduated in degrees and minutes) with a radius of about 430 centimetres. Equipped with an alidade of two sights, it was used to determine the altitude of heavenly bodies. With it Tusi's astronomers determined the latitude of Maragha as well as the obliquity of the ecliptic. The armillary sphere or spherical astrolabe had five rings (the radius of the outer ring was 160 centimetres) and an alidade. It was used primarily to measure the longitude and latitude of the ecliptic. A solstitial armilla with an alidade and a ring (250 centimetres in diameter) was used to calculate the obliquity of the ecliptic. An equinoctial armilla determined the Sun's path at the vernal and autumnal equinoxes and at the moment of its entry into the equatorial plane. A dioptrical ruler of Hipparchus, a two-holed instrument, measured the apparent diameters of the Sun and Moon and was also used to observe solar and lunar eclipses. An azimuth ring with two alidades gauged the angle of elevation of heavenly bodies (in degrees and minutes). This instrument, although designed by 'Urdi, was not ready for immediate use. In his treatise he referred only to a model, and the first eye-witness accounts date to the next century. A parallactic ruler (with a radius of 250 centimetres) was used to measure the altitude of heavenly bodies. Two azimuth instruments determined the sine and versine of the angles of elevation of heavenly bodies. The fact that 'Urdi planned but did not construct all of these instruments suggests that his stay at Maragha was relatively short and that there were other skilled instrument makers on Tusi's staff. In addition to the relatively large and immobile instruments described in 'Urdi's treatise, there were also a large number of smaller portable devices in the possession of individual astronomers.[20]

Since the astronomical instruments were out of doors or in the small domed annex, the large central building was the heart of the institution – at once library, school, office complex, and administrative centre. Hulagu and Nasir al-Din had assembled an impressive collection of books – as many as 40,000. The result of Mongol conquests of Baghdad and Syria, this was the first major observatory library in the Islamic world and was partly responsible for the number and quality of the scientific works produced by the staff. Although we don't know the names of all of the Maragha astronomers, the size and wealth of the institution and the fact that it remained active for more than fifty years suggests that the total number was relatively large – seventy-five to a hundred, perhaps. The sources mention fifteen or twenty of the most prominent. Tusi himself and 'Urdi but in addition 'Ali ibn 'Umar al-Qazwini, Fakhr al-Din al-Akhlati, Fakhr al-Din al-Maraghi, Muhy al-Din al-Maghribi, Qutb al-Din al-Shirazi, Shams al-Din al-Shirwani, Najm al-Din Dabiran al-Qazwini, 'Abd al Razzaq ibn al-Fuwati, the librarian, and Kamal al-Din al-Ayki. 'Urdi, Akhlati, Maraghi, and Qazvini were among the original group recruited by Tusi and were probably the senior members of the staff. Although Qutb al-Din al-Shirazi had been an early student of Tusi, by the time of Hulagu's death in 1265 he had become a full-fledged member of the institution. A Chinese astronomer, Fao Mun Ji, provided information about Chinese astronomy and astrology. In addition to the scientists, there must have been a large body of support personnel – technicians, servants, students, and labourers.[21]

The Maragha Observatory was the most advanced and productive scientific institution of its day. Its superiority was due to three factors: financing, longevity, and educational activity. Hulagu was the first ruler to endow his observatory with a *waqf* (a permanent and inalienable source of income). At this time most *waqf*s supported charitable or public institutions (mosques, madrasas, or hospitals), and Hulagu was at first reluctant to depart from this tradition. As the observatory grew in size and expense, however, Tusi had to make repeated requests for funds. Finally, he convinced the ruler to establish a permanent endowment – said to total ten percent of the entire Ilkhanate charitable budget. Hulagu appointed Tusi administrator, not just of the observatory endowment, but of the entire *waqf* department.[22]

Hulagu's observatory was also unusual in its lifespan. Up to this point most of the observatories in the Islamic world had not outlived their founders. At Maragha, however, the resident astronomers continued to make observations, calculate planetary positions, prepare almanacs, and compile astronomical treatises for the next half century or so. The observatory remained a productive and functioning institution throughout the reigns of the remaining rulers of the dynasty – from Abaqa (1265–81), Hulagu's successor, through Uljaytu (1303–16), the penultimate Ilkhanate monarch. Two of Tusi's sons succeeded him as director of the observatory – first Sadr al-Din and then 'Asil al-Din, who was appointed by Uljaytu in 1304 and remained in his post until his death in 1316. The buildings and instruments survived the fall of the Ilkhanate Dynasty. Ulugh Beg (1394–1449), the Timurid ruler, toured Maragha during his childhood, and the mathematician Jamshid al-Kashi (1380–1429) saw a geometrical pulpit during his visit. At the beginning of his reign Shah Isma'il I (r. 1501–24), the founder of the Safavid dynasty, considered renovating the observatory. And as late as 1562–3 a traveller reported that the wall of the large mural quadrant was still standing.[23]

The Maragha Observatory was also an educational institution. Since most madrasas offered instruction in the traditional sciences only, there were not many institutionalised settings for the study of the rational sciences. And for astronomy and astrology, in particular, theoretical knowledge needed to be supplemented by practical, hands-on-experience with instruments and calculations. Thus, from the very beginning organised instruction was an important part of the institution's mission. Abu Faraj, one of the early members of Tusi's staff, seems to have specialised in teaching rather than research. From 1272 to 1279 he lectured on geometry, mathematics, and astronomy – covering Euclid's *Elements* in 1270 and Ptolemy's *Almagest* in 1272. His lecture notes drew heavily on the writings of Tusi himself. The observatory provided support for students as well as for professors. Abaqa, Hulagu's successor, awarded stipends to nearly one hundred of Tusi's students after his death.[24]

The principal achievement of the Maragha astronomers was the *Zij-i Ilkhani*. An astrological document primarily, this treatise updated the various planetary parameters, providing the practicing astrologer with the best available information for casting horoscopes. Hulagu, like most of his contemporaries, was heavily dependent on the advice of astrologers. He never took a trip or made a significant decision without first consulting Tusi, and on his death Abaqa would not ascend the throne until an auspicious time had been determined. After the death of Ghazan Khan, Uljaytu likewise delayed his accession until the astrologers had finished their deliberations. Given Tusi's reputation as an expert in the rational sciences, it is ironic that the best-known creation of his observatory was dedicated to a superstitious rather than a scientific end.[25]

Composed in Persian by Tusi himself, the *Zij-i Ilkhani* was begun in 1259, when the Maragha Observatory was founded, and finished in 1271–2. Although the text mentioned earlier treatises (those of Ibn Mansur, c. 810; Ibn Yunus, c. 900; and al-Battani, c. 900), Tusi's treatise was an immediate success, superseding all earlier works after it was translated into Arabic. For over 150 years, until the *Zij al-Sultani* of Ulugh Beg, Tusi's work was the indispensable reference for the astronomer/astrologers of the Islamic world.

The *Zij-i Ilkhani* was divided into the usual categories. An opening section on chronology described the Hijra, Yazdegird, Seleucid, Jewish, Maliki (or Jalali), and Chinese 12-Year Animal (evidence of the Far Eastern influence) calendars and eras. In keeping with Tusi's contribution to the field, two long sections were devoted to trigonometry. The first contained sine tables to three places, and the second, on spherical trigonometry, had planetary ascension tables to two places. Another table, not seen in earlier treatises, displayed ascension and equatorial angles for every hour (from sunrise to sunset) given in seconds of arc. There were equations for determining the true solar time, correcting for the differences between mean and apparent solar time. The mean positions of all planets (computed to seconds of arc) were given for the years 600–700 of the Yazdegird Era. Tables of planetary equations and planetary latitudes were computed to arcs of twelve minutes each. Another table listed the zodiacal stations for each planet. Planetary sectors – of deferent and epicycle motions – were computed to degrees of arc. A parallax table corrected for the difference between the apparent position of a planet and its position as computed on a geocentric basis. There were tables listing solar and lunar eclipses – with their magnitudes and Hijra dates. A visibility table displayed the dates for the first appearance of the Moon and the planets. Tusi also put together a lengthy table giving latitudes and longitudes in five minute intervals for 245 localities. The length of the day was given for the thirty-five largest cities. The Maragha astronomers also produced (from their own observations) a new star table: the latitude, longitude, temperament, and magnitude of sixty stars. The final group of tables was more exclusively astrological, giving the planetary positions for the various signs of the zodiac and providing information on world years and planetary conjunctions, especially of Jupiter and Saturn.[26]

Like the Seljuq ruler Jalal al-Din Malik Shah, Hulagu was deeply disappointed when told that it would take thirty years to compile a completely new astronomical treatise. Unlike 'Umar Khayyam, however, Nasir al-Din Tusi did not abandon the effort, promising the Ilkhanate ruler that he would expedite the process. And Tusi did in fact finish the *Zij-i Ilkhani* in 1272, some thirteen years after the observatory was founded, but to no avail. Hulagu died in 1265.

In contrast to the Seljuq observatory in Isfahan, the Ilkhanate observatory in Maragha remained active for more than fifty years, time enough to complete a full cycle of planetary observations. Whether this was actually done, however, is unclear. The fifteenth century astronomer Rukn al-Din ibn Sharaf al-Din al-Amuli wrote in his *Zij-i Jami-i Sa'idi*:

> As is well known Tusi . . . had made certain mistakes in the Ilkhani Zij and had willed that these mistakes be rectified and the tables in the zij be corrected by Asil al Din in collaboration with Qutb al Din al Shirazi. Now the Khwaja (i.e. Tusi) had mentioned the names of the astronomers of the observatory in the introduction to the Ilkhani Tables and had passed away, and had not included the Mawlawi's (Qutb al Din) name among them. Because of this Qutb al Din did not busy himself with the correction of the tables. Upon Asil al Din's insistence he merely indicated on the margins . . . that in using the mean positions of the planets from the tables, 30 minutes should be added to the mean position of the moon and 7 minutes to the centre of Saturn's epicycle, that Jupiter's epicyclic configuration should be increased by 1 degree 21 minutes and that 1 degree 30 minutes should be added to the centre of Mars's epicycle and the same quantity subtracted from that of Venus, but he did not make any references to the sun and to Mercury.
>
> After the death of the Khwaja, the astronomers of the observatory waited up to thirty years until the revolution of Saturn became complete. Each one of them, such as Athir al Din al Abhar, Muhyi al Din al Maghribi, Najm al Din Dabiran, and Fakhr al Din al Akhlati, prepared astronomical tables and treatises on Euclid and the Almagest, and it was ascertained that, as before, three minutes should be subtracted from the sun's distance to the apogee of its eccentric so as to bring about the necessary agreement between the calculated and observed positions.[27]

A later Maragha astronomer, however, told a slightly different story. Muhammad ibn 'Ali Khwaja Shams al-Munajjim al-Wabkanwi remained at the observatory for more than forty years, dedicating his treatise (*Zij-i Shamsi-i Munajjim*, c. 1320) to Abu Sa'id Bahadur Khan (1316–36).[28] Wabkanwi stated that the Ilkhani tables were not accurate because they were not based on a thirty-year program. According to him, it was Muhy al-Din Maghribi (1220–83), a senior member of the Maragha staff, who completed the original programme of observations.

Of the two Maragha astronomers mentioned above Qutb al-Din Shirazi (1236–1311) was the younger. Born in Shiraz, he studied medicine and mysticism under his father, a famous physician and Sufi. Like the other polymaths he was precocious. His father died when he was fourteen, and he was chosen to succeed him at the local hospital. He remained in Shiraz for ten more years, studying the definitive medical text of his day – Ibn Sina's *Canon of Medicine*. In 1260, however, dissatisfied with his progress, he left Shiraz in pursuit of a teacher who could more fully explain Ibn Sin's medical and philosophical theories. Two

years later he arrived in Maragha, drawn by Tusi's reputation as a commentator on Ibn Sina. Once under the master's sway, however, Shirazi expanded his
focus, adding astronomy and mathematics to philosophy and medicine. Shirazi
remained in Maragha for another decade or so and became the observatory's
most famous and most productive student. Sometime before Tusi's death in
1274, Shirazi left Maragha and resumed his scholarly journey – spending time
in, among others, Qazvin and Baghdad, before ending his days in Tabriz.

Like the other polymaths, Shirazi was prolific. He wrote definitive works
in philosophy, theology, optics, geography, astronomy, and medicine. His
five-volume commentary on Ibn Sina's *Canon*, which he worked on for most
of his life, was perhaps his most famous work but, for our purposes, his theoretical work on planetary motion was the most important. Although Shirazi
participated in the work of the Maragha Observatory during his student days,
he probably did not play a major role in the preparation of the *Zij-i Ilkhani* (thus
his absence from its list of astronomers and his perfunctory corrections for Asil
al-Din). Shirazi's contributions to the work of the Maragha School were not
completed until well after he left the Observatory. In 1282 he completed *The
Limit of Understanding the Knowledge of the Heavens*, his major work; in 1284 he finished *The Royal Gift on Astronomy*; and in c. 1304 he compiled *Muzaffari Selections*.[29]
In these treatises, Shirazi constructed prototypes of planetary motion that eliminated the Ptolemaic equant. Employing the 'Tusi Couple', he produced models
that were both mathematically accurate and consistent with Aristotelean laws
of physics. Although Shirazi's theories were later refined and perfected by Ibn
Shatir, he was a significant influence on Regiomontanus and to a lesser degree
on Copernicus himself.

The other prominent astronomer of the Maragha School, in addition
to Shirazi and 'Urdi, was Muhy al-Din al-Maghribi (c. 1220–83). Born in
Andalusia, he acquired a thorough grounding in mathematics and astronomy,
probably from the astronomers of the Toledo School. At some point, perhaps in
the late 1240s or the early 1250s, he arrived in Damascus.[30] The only evidence
of his sojourn in the Syrian capital was a *zij*, *The Crown of the Astronomical Handbooks
and the Satisfaction of the Needy*.[31] This treatise was completed in 1258, an uncertain
year for Damascus and Muhy al-Din. Hulagu had just conquered Baghdad
and Damascus would fall two years later. In the chronological section of his *zij*
Muhy al-Din included the era of Hulagu, suggesting that he may have planned
to dedicate his treatise to the Mongol ruler. In any event, the astronomer was an
early invitee and became one of the most talented and productive of Tusi's collaborators. He remained at the observatory for twenty-five years – assisting Tusi
and 'Urdi in the construction of the observatory, making a number of observations between 1262 and 1274, and, given his experience, probably playing a
major role in the drafting of the *Zij-i Ilkhani*. He also composed a number of
important mathematical and astronomical works. In trigonometry, for example,

he calculated the value of the sine of one degree to an accuracy of four decimal places, and he arrived at one of the most accurate estimates for π.

In 1280, toward the end of his life, al-Maghribi compiled another treatise: *Zīj-i Muhy al-Dīn al-Maghribi al-Andalus*.[32] This is undoubtedly the one mentioned by Wabkanwi, and although it must have been more accurate than the *Zīj-i Ilkhani*, being based on another decade or so of work, it could not have been complete. An entirely new set of tables would have taken a full thirty years (the period of Saturn's revolution). If begun in 1259, the treatise could not have been finished until 1289, and if begun in 1274 (the year of Tusi's death) it cannot have been completed until 1304. And Maghribi unfortunately died in 1283. A final possibility, as author of a completely new and accurate *zīj*, was Wabkanwi himself. Like Tusi and Maghribi, he compiled a treatise (dated 1320) and unlike them he remained a practicing member of the observatory for more than thirty years. The ultimate answer to these questions about Tusi's *zīj* is at present beyond us. Until a careful analysis of the various treatises – the Ilkhani in all its versions along with those of al-Wabkanwi and al-Maghribi – is undertaken, it is impossible to finally assess the astronomical and mathematical achievements of the Maragha Observatory.[33]

After its translation into Arabic, Tusi's *zīj* quickly swept aside the older treatises, garnering an impressive number of translations and commentaries. An otherwise forgettable commentary by Muhammad Shah Khulji survived in a 1652 translation into Latin by John Greaves, the Oxford professor of astronomy.[34] And in the thirteenth and fourteenth centuries parts of the *zīj* were translated into Byzantine Greek by Gregory Chioniades (c. 1320) and George Chrysococces (c. 1350).[35] Tusi also influenced the early European astronomers. In 1496 Regiomontanus (1436–96), the German mathematician, astronomer, and astrologer, published an epitome of the *Almagest*. In it he employed some of the Tusi's techniques for correcting and simplifying the complicated epicyclic system of Ptolemy's planetary model. And Nicholas Copernicus used the 'Tusi Couple' in his early *Commentariolus* and in his path breaking *De revolutionibus*. One suggestion is that he may have picked up the idea from Regiomontanus' epitome of the *Almagest*.[36]

Just as the comparison of the treatises of 'Umar Khayyam (*Zīj-i Malik Shahi*) and al-Khazini (*Zīj al-Mu'tabar al-Sanjari*) with the *Toledan Tables* of al-Zarqali revealed that in the late eleventh century the forefront of Islamic astronomy was in the East, so too in the late thirteenth century. The *Zīj-i Ilkhani* of Nasir al-Din Tusi and the Castilian canons of the *Alfonsine Tables* were exactly contemporaneous – 1271. But a comparison of the two astronomical treatises reveals that the gap between the two halves of the Islamic world had widened even further. Alfonso X's *zīj* was compiled by two men working for about nine years with small portable instruments. They made very few new observations and the Castilian version of the *Alfonsine Tables* appears to have been mostly derivative – modelled after al-Majriti's adaptation of the *Zij al-Sindhind* with additional material from al-Battani and Ptolemy. Nasir al-Din Tusi's *zij*, on the other hand, in its several versions, was the work of an entire school of astronomers working for more than fifty years in an observatory that boasted the largest and most accurate collection of astronomical instruments in the Eurasian world. The treatise, or more accurately, the treatises that resulted represented the most sophisticated mathematical and astronomical achievements in the Eurasian world. The 'Tusi Couple' and the "Urdi Lemma' allowed the Maragha astronomers to eliminate some of the worst contradictions of the Ptolemaic model and to produce models of planetary motion that were consistent, both with the laws of physics and the predictions of mathematics. Because, however, the breakthroughs in Maragha – the 'new astronomy', according to Ibn al-Shatir of Damascus (d. 1375) – occurred after the Reconquista in Andalusia, it would be another two hundred years before the discoveries of the Maragha School became fully known in Renaissance Europe.

Notes

1. For an overview see Hockey, *Biographical Encyclopedia of Astronomers*, 1153–5; 'al-Tusi, Muhammad Ibn Muhammad ibn al-Hasan, usually known as Nasir al-Din', *Dictionary of Scientific Biography*; 'Tusi, Nasir al-Din', *Encyclopaedia Iranica*. See also North, *Cosmos*, 205–7; F. Jami Ragep, 'Tusi and Copernicus: The Earth's Motion in Context', *Science in Context* 14 (2001): 145–63.
2. 'al-Tusi', *Dictionary of Scientific Biography*; 'Aklaq-e Naseri', *Encyclopaedia Iranica*; G. M. Wickens (tr.), *Akhlaq-i Nasiri* (London: George Allen and Unwin, 1964).
3. North, *Cosmos*, 205–7; 'Tusi, Nasir al-Din', *Encyclopaedia Iranica*; O. Gingerich, 'A Tusi Couple from Schoener's "De Revolutionibus"', *Journal of the History of Astronomy* 15 (2) (1984): 128–33; J. F. Ragep, 'Two Versions of the Tusi Couple', in *From Deferent to Equant* (New York, 1987), 329–56.
4. 'al-Tusi', *Dictionary of Scientific Biography*.
5. Jami Ragep, *Nasir al Din Tusi's Memoir on Astronomy*, 2 vols (New York: Springer-Verlag, 1993).
6. Ibid.
7. 'Tusi, Nasir al-Din', *Encyclopaedia Iranica*.
8. Ibid.
9. 'al-Tusi', *Dictionary of Scientific Biography*.
10. Ibid.
11. Ibid.
12. Ibid.
13. Sayili, *Observatory*, 187–8.
14. Ibid. 189.
15. Ibid. 39.
16. Ibid. 202–3.
17. Hockey, *Biographical Encyclopedia of Astronomers*, 1161–2.
18. Sayili, *Observatory*, 193–4.
19. Ibid. 195.

20. Ibid. 199–201. For a translation of 'Urdi's *Treatise on Instrumentation* see Prof. Sevim Tekkeli, "Urdi's article of "The Quality of Observation"', in *Foundation for Science, Technology and Civilisation* (Manchester, 2007).
21. Sayili, *Observatory*, 205–7.
22. Ibid. 207–10.
23. Ibid. 211–14.
24. Ibid. 218–22.
25. Ibid. 203–4.
26. Kennedy, *Islamic Astronomical Tables*, 161–2.
27. Sayili, *Observatory*, 213–14.
28. Kennedy, *Islamic Astronomical Tables*, 137.
29. 'Shirazi: Qutb al Din Mahmud ibn Masud Muslih al Shirazi', in Hockey, *Biographical Encyclopedia of Astronomers*, 1054–5; 'Qutb al Din al Shirazi', *Dictionary of Scientific Biography*.
30. 'Muhyi 'l Din al-Maghribi', *Dictionary of Scientific Biography*.
31. Carlos Dorce, 'The Taj al-azyaj of Muhyi al-Din al-Maghribi (d. 1283): Methods of Computation', *Suhay* 3 (2002–3): 193–212.
32. Kennedy, *Islamic Astronomical Tables*, 131.
33. George Saliba, 'Solar Observations at the Maragah Observatory Before 1275: A New Set of Parameters', in Saliba, *A History of Arabic Astronomy*, 177–86.
34. Kennedy, *Islamic Astronomical Tables*, 161–2.
35. 'Tusi, Nasir al-Din', *Encyclopaedia Iranica*.
36. North, *Cosmos*, 206–7.

CHAPTER SIX

The observatory in Samarqand

The century and a half which separated Nasir al-Din's observatory in Maragha (1262) from the one constructed by Ulugh Beg in Samarqand (1420) was not without interest. In 1300 the Ilkhanid ruler Ghazan Khan (1295–1304) founded a complex of institutions outside his capital of Tabriz. At the centre of this development was a mausoleum, around which he erected a mosque, a Sufi lodge (*khanqah*), two madrasas, a hospital, a library, a law school, a primary school, a bathhouse, and an observatory. Like Hulagu, Ghazan Khan was interested in both the science and the pseudoscience, and, after visiting Maragha, he designed an observatory which contained a hemispherical instrument for solar observations, a library, and a school for teaching the rational sciences. Ghazan Khan's Tabriz Observatory, however, was much smaller than the one at Maragha. With a limited programme and a shorter lifespan, its principal achievement was the creation of a new calendar and a new era. The epoch of the Khani Era was 1302, and its calendar was quite similar to the Jalali of 'Umar Khayyam. The year began on the vernal equinox (21 March) but the names of the Khani months were Turkish rather than Persian. The new era and calendar, however, were not widely adopted, remaining in use only through the reign of the last Ilkhanid ruler, Abu Sa'id Bahadur (1316–36).[1]

In 1325 Rukn al-Din Ahmad, a pious *sayyid*, founded in the Iranian city of Yazd a charitable complex consisting of a madrasa, a mosque, a hospital, a library, and what a local historian called the *Rasad-i Waqt wa Sa'at* (*The Observatory of the Time and Hour*). Although 'observatory' suggests a building or buildings, observational instruments, and astronomers, where the movements of the heavenly bodies were tracked, recorded, and interpreted, the actual description (and the name – 'time and hour') indicates more a *muwaqqit khana* (timekeeper's office) than an astronomical institution. From the description in *Tarikh-i Kabir* (a local history), the structure seems to have been a giant astronomical water clock. Constructed in a tall tower at one side of the madrasa, the water clock was a complex arrangement of wheels, buckets, and balls. As the water levels changed the clock displayed the names of the months (in the Hijra, Jalali, Rumi, and Christian calendars), the days of the month, the hours of the day, the mansions of the Moon, and the positions of the five planets. Built during the reign of the last Ilkhanid ruler, Abu Sa'id Bahadur (1316–36), the observatory seems to have been another example of the Ilkhanid interest in the heavenly sciences.[2]

The most important astronomer of the fourteenth century was Ibn al-Shatir (1304–75). He bridged the century between the scientists of the Maragha Observatory (Tusi, 'Urdi, and Shirazi) and those of the Samarqand Observatory (Ulugh Beg, Ghiyath al-Din Jamshid al-Kashi, Qadi Zada al-Rumi, and 'Ala al-Din 'Ali al-Qushji). He was the chief timekeeper of the Umayyad Mosque in Damascus and was said to have founded his own observatory. Although there is no contemporary evidence of an actual structure, it is certain that he organised a programme of observations, composed treatises on planetary theory, and compiled an astronomical treatise.[3]

Finding the instruments of his day difficult to build and operate, Ibn Shatir constructed a universal astrolabe and an azimuthal quadrant and drew up plans for a large horizontal sundial on the northern minaret of the Umayyad Mosque. He also authored works on the planispheric astrolabe and the astrolabic and trigonometric quadrants.[4] His most important contribution, however, was in planetary theory. In *The Final Quest Concerning the Rectification of Principles* Ibn Shatir drastically refashioned the Ptolemaic models for the Sun, Moon, and planets. The Damascus astronomer utilised the 'Tusi Couple' and added new epicycles to the Ptolemaic construct. He fashioned a system that eliminated the epicycle in the solar model, the eccentrics and equants in the planetary models, and the eccentric, epicycles, and equant in the lunar model. His new system was one in which the Earth was at the exact centre of the universe, and the heavenly bodies revolved around it in uniform circular motions. Ibn Shatir was as accurate as Ptolemy in predicting the paths of the heavenly bodies but he improved on the Greek by accurately predicting the apparent size of the Sun and Moon and their distance from the Earth. Ibn Shatir also seems to have influenced the heliocentric model of Nicolas Copernicus. Although the exact path of transmission remains unknown, Copernicus's models for the Moon and Mercury were identical to those of the fourteenth-century Islamic astronomer.[5]

Ibn Shatir also compiled an astronomical treatise, *al-Zij al-Jadid* (*The New Astronomical Treatise*). Finished in about 1350 and based on observations made in Damascus, this was the most accurate and complete set of tables between the *Zij-i Ilkhani* of Nasir al-Din Tusi and the *Zij al-Sultani* of Ulugh Beg. The first section on chronology included descriptions of the Hijra, Yazdegird, Seleucid, and Coptic eras and calendars. Tables of trigonometric and spherical astronomical functions were computed to four places – one table contained 184 identities to four terms. Equations of time and mean motions for both the inferior and superior planets were given for the Hijra era at 30, 60, 90, . . . 900 years and for the appropriate months, days, hours, and minutes at Hijra years 1, 2, 3 . . . 30 to thirds of an arc. Precessional motion was tabulated to months and days and to four fractional places. Planetary equations and latitudes were found along with a table of stations computed to seconds of arc. Ibn Shatir reproduced Ptolemy's tables for solar and lunar parallax, but with an interpolation scheme that provided the results for Damascus. These tables were divided for each minute of arc after sunrise, and the conjunctions were given for the midpoint of each zodiacal sign. He also included the tables from the *Zij al-Sindhind* of al-Khwarizmi – with full and accurate directions. A table for eclipses was calculated to two places for each degree. There were two separate visibility tables for the Moon along with a table at intervals of three degrees for the difference between rising and setting for all seven climes. The visibility tables for the planets were taken from the *Almagest* – with an interpolation scheme appended. Coordinates were furnished for eighty stars along with longitude, latitude, and *qibla* directions in minutes for two hundred cities. The *zij* ended with an astrological table – a list of the lots used to determine various outcomes at the vernal equinox.[6]

In 1420 the Timurid prince Ulugh Beg (1394–1449) began work on the largest and most sophisticated observatory in the Eurasian world. Like the earlier founders of major institutions (Jalal al-Din Malik Shah and Hulagu), Ulugh Beg also relied on the forecasts of astrologers but, unlike them, he harboured a passion for the rational sciences – mathematics and astronomy especially. After the death of his grandfather, the great Central Asian conqueror Timur (1336–1405), Ulugh Beg followed his father, Shah Rukh (1405–47), ruler of the eastern half of the Timurid Empire, to Samarqand. In 1409 Shah Rukh transferred his capital to Herat, and in 1410 he appointed his sixteen-year-old son governor of Turkestan and Transoxiana. Content with a subordinate political role, Ulugh Beg devoted himself to the arts of peace – founding a madrasa and an observatory in Samarqand and underwriting religious and cultural, as well as educational and scientific activities. After Shah Rukh's death in 1447, Ulugh Beg briefly ascended to his father's throne. Lacking the knack for political intrigue, however, he was easily outmanoeuvred by his nephew, and two years later was defeated and beheaded by his son.[7]

Ulugh Beg was an intellectual prodigy, precociously talented in mathematics and astronomy, but bored by war and government. Of the prince Ghiyath al-Din Jamshid al-Kashi (c. 1380–1429), the famous astronomer and mathematician, wrote:

the king of Islam . . . is a learned person . . .[8]

. . . [he] . . . is an astronomer, the director of the observatory and . . . is actively engaged in this work . . .[9]

[he] . . . has great skill in the branches of mathematics. His accomplishment in these matters has reached such a degree that one day, while riding, he wished to find out what day of the solar year a certain date would correspond which was known to be a Monday of the month of Rajab in the year 818 and falling between the tenth and the fifteenth of the month. On the basis of these data he derived the longitude of the sun to a fraction of two minutes by mental calculation while riding on horseback . . .[10]

Solving complicated astronomical equations in his head was not Ulugh Beg's only skill, he also had a photographic memory. Not only had he memorised most of the Qur'an, but he had complete recall in other areas as well. An enthusiastic hunter, he kept a detailed diary of the birds he had shot – names, dates, and places. One day his librarian reported the book lost. Ulugh Beg told him not to worry and promptly dictated from memory the contents of the entire volume.[11]

During his first years in Samarqand (1410–17) Ulugh Beg organised a circle of like-minded students under the direction of the Ottoman mathematician-astronomer Salah al-Din Musa Pasha, known as Qadi Zada al-Rumi (c. 1359–1440). Born in Bursa, capital of the early Ottoman Empire, Qadi Zada was the son and grandson of *qadi*s (judges of Islamic law). His early education was un-traditional. In addition to the time-honoured traditional sciences, he also studied the rational sciences with Muhammad al-Fanari (d. 1431). Remaining in Bursa for the first half of his life, he began to specialise in mathematics and astronomy, writing in 1383 a treatise, *Introduction to Arithmetic,* that later became a standard Ottoman textbook. After al-Fanari suggested that he needed further study, Qadi Zada travelled to the intellectual centres of the eastern Islamic world. Arriving in Timur's Samarqand toward the end of the fourteenth century, he spent several years studying with Sayyid al-Jurjani (1340–1413). Although Jurjani had written commentaries and super-commentaries (commentaries on commentar-ies) on several classic works in astronomy (those of Tusi, Shirazi, and Jaghmini), Qadi Zada decided that he was deficient in mathematics. On Timur's death in 1405, Jurjani returned to Shiraz and Qadi Zada continued his intellectual wandering. Although his exact itinerary is unknown, he probably spent some time in both Herat and Bukhara. In any event, in c. 1412 he resurfaced in Samarqand. By this time a mature scholar and teacher of fifty-two, Qadi Zada was to remain in Ulugh Beg's capital for the rest of his life. He married, fathered a son (Shams al-Din Muhammad), died, and was buried outside the city walls, near the mausoleum of the famous Shah-i Zinda.

In 1412 Ulugh Beg, the seventeen-year-old, newly-appointed governor of Khurasan and Transoxiana, asked Qadi Zada to lead a seminar devoted to mathematics and astronomy. The Ottoman scholar directed this informal colloquium for the next five years, and his writing during this period reflected both his appointment and his intellectual bent – he was more a teacher and commentator than a theoretician and innovator. In 1412, during his first year in Samarqand, Qadi Zada wrote two commentaries. The first, on Mahmud ibn Jaghmini's (d. c. 1221) *A Compendium on Astronomy,* was dedicated to Ulugh Beg. Jaghmini's work, a popular elementary text, became even more widely known in Qadi Zada's recension. Itself the subject of many super-commentaries, this commentary eventually became an Ottoman standard. The second com-mentary, also dedicated to Ulugh Beg, took up Shams al-Din Samarqandi's (c. 1250–1310) *The Fundamental Theorems.* Samarqandi's treatise was short – twenty

pages on thirty-five of Euclid's theorems – and very popular. Qadi Zada's treatise, also widely admired, became *the* Ottoman textbook on geometry. It was the
subject of many super-commentaries and was translated into Ottoman Turkish
in the nineteenth century.

In 1417 Ulugh Beg broke ground on a magnificent new madrasa on the
central square of Samarqand. The institution specialised in the rational sciences,
and Qadi Zada, as its head, was responsible for the lectures on mathematics and
astronomy. Over the next three years the madrasa grew in size and importance,
attracting talented scholar-teachers and ambitious students. Ulugh Beg was
himself an exceptionally precocious early student and attended both the lectures
and the informal discussions that followed. He did not, however, dominate
the conversations. In the young prince's mind, there was no room in scientific
matters for courtesy; he debated at length with the other students and was impa-
tient with those who agreed with him merely out of politeness. Later the prince
joined the faculty, giving lectures whenever time allowed.[12]

In 1420 work began on the new observatory. As the institution became
operable more and more of the madrasa astronomers joined its staff. Qadi
Zada became, along with Jamshid al-Kashi and Ulugh Beg, a director of the
new observatory, and he soon began to participate in its day to day operation
– manning equipment, collecting data, and working on the first draft of the *Zij
al-Sultani*. Unlike the other astronomers, however, Qadi Zada's attention was
divided. In the observatory's first decade his primary responsibility was the
madrasa. After al-Kashi died in 1430, however, Ulugh Beg asked him to devote
more time to the observatory and during the following ten years he shuttled
back and forth between the two institutions.

During his twenty odd years in Samarqand, Qadi Zada continued to write.
He completed a commentary on Tusi's famous criticism of Ptolemy (*A Treatise
on the Science of Astronomy*) and a super-commentary on Ahmad ibn Mahmud
al-Harawi's commentary on Athir al-Din al-Abhari's (c. 1200–64) *A Guide to
Philosophy*. This last, on logic, physics, and metaphysics, also became a madrasa
standard. Although Qadi Zada's writings were mostly in Arabic (the language
of Islamic science), he also composed two works in Persian: a treatise on finding
the *qibla* direction and a work for revenue officials on surveying. Two of his later
compositions were more original. His treatise on the calculation of the sine of
one degree, a commentary on al-Kashi's famous *Treatise on the Chord and Sine*,
introduced a new method of calculation, simplifying and extending al-Kashi's
approach. And he was largely responsible for the new star catalogue in the *Zij
al-Sultani*, finally finished during the last year of his life.[13]

Ulugh Beg had visited the ruins of Maragha as a child and had discovered
during his madrasa studies that the *Zij-i Ilkhani* of Nasi al-Din Tusi was badly
out of date. As a result, he decided to establish an observatory and to compile a
new and more accurate treatise. Of the Samarqand Observatory, Zahir al-Din

Muhammad Babur (1483–1530), founder of the Mughal mpire (1526–1739) in India, wrote: '. . . the observatory . . . is the instrument for the writing of astronomical tables; it is three stories high. Ulugh Beg Mirza prepared the Zij-i Gurgani (or Sultani) at this observatory . . .'[14]

To found such an institution, however, wealth was not enough. Hasan ibn Muhammad Nizam al-Nishapuri, author of *Sharh-i Zij-i Ilkhani* (*Commentary on the Zij-i Ilkhani*), wrote:

> . . . This [erection of an observatory] is not merely due to the fact that great expenditures are necessary . . . For there is need for the presence of accomplished and skillful masters who can convert the instruments from the conceptual state into the actual . . . It is necessary . . . to bring together all the masters of the time so that every one shall make manifest his own particular art . . . Now, there is no doubt that to assemble the masters of this art from all corners is possible sometimes by showing kindness to them, and in other cases through compulsion and harshness; but kindness is more effective when it comes from kings and coercion can be exercised only by them . . .[15]

By the time Ulugh Beg founded his madrasa and built his observatory, astronomy had become very specialised. There were astronomers who had theoretical knowledge but lacked any experience in making observations and recording and calculating the results. On the other hand, there were also astronomers who knew only the practical side of the field: they made observations and plugged the measured quantities into the relevant formulas without much understanding of the theories involved. The titles given to the men of the Samarqand Observatory illustrate the point. There was the *riyadi* (mathematician), the *hasib* (calculator), the *muhandis* (engineer), the *handasi* (the geometrician), the *'adadi* (arithmetician), the *falaki* (astronomer), the *munajjim* (astrologer), the *rasid* (observer), and the *usturlabi* (instrument designer). And the specialities were sometimes combined – yielding an *al-riyadi al-muhandis* (a mathematician and engineer), an *al-hasib al-falaki* (a calculator and astronomer), an *al-riyadi al-falaki* (a mathematician and astronomer), and an *al-hasib al-muhandis* (calculator and engineer).[16]

Ulugh Beg was the director of the Observatory and Qadi Zada al-Rumi was the head of the Madrasa but the intellectual star, the leading mathematician and astronomer, was Ghiyath al-Din Jamshid al-Kashi. By the time he arrived at the Samarqand Madrasa in about 1417–18, al-Kashi was a mature scholar (37 years old) with a long list of publications and a sterling reputation. To his contemporaries he was the 'support of astronomy', the 'second Ptolemy', the 'master teacher of the world'.

About al-Kashi's early life and education very little is known. He grew up in the town of Kashan in north central Iran. His early training in the rational sciences (including medicine) was in Kashan, but like al-Biruni, Tusi, and Qadi

Zada himself he had to leave his hometown in search of opportunity and liveli-
hood. Al-Kashi's odyssey before landing in Samarqand is uncertain, its outline
suggested only by the dates and places of his writings. In 1406 he observed an
eclipse in Kashan. In 1407, still in Kashan, he dedicated *The Stairway of Heaven* to
the otherwise unknown vizier, Kamal al-Din. This was an attempt to resolve the
difficulties of his predecessors (especially Ptolemy) in determining the relative
size and distance of the heavenly bodies. In 1410–11 he wrote *A Compendium of
the Science of Astronomy* for Sultan Iskandar, a cousin of Ulugh Beg, who ruled the
Iranian provinces of Fars and Isfahan. A comprehensive overview, it probably
served as a general introduction for both his madrasa and observatory students.
In 1414 in Samarqand he dedicated his *Zij-i Khaqani* (*The Astronomical Treatise of
the Khan*) to Ulugh Beg. At this time al-Kashi probably addressed several sessions
of Qadi Zada's discussion group. In 1416 in Tabriz he dedicated *A Treatise on
the Explanation of Observational Instruments* to Sultan Iskandar (1420–36), the Qara
Qoyunlu (Black Sheep) ruler. Later in the same year, back in Kashan, he completed *The Garden Excursion*, another work on astronomical instruments. In 1417
or 1418 al-Kashi returned to Samarqand, joining the faculty of the madrasa,
probably at the invitation of either Qadi Zada or Ulugh Beg or both.[17]

The most important of al-Kashi's early writings (as least for his career in
Samarqand) was the *Zij-i Khaqani*. It was an effort to correct the mistakes of
Tusi's *Zij-i Ilkhani*. Although al-Kashi had no access to the large instruments
of an observatory and could make no new observations (except for three solar
eclipses), he was able to take advantage of the theoretical and observational
advances of the previous century and half to update or revise many of Tusi's
tables. Each of the six sections of the *zij* was organised into three parts: an extensive technical glossary, an explanation of operations, and a brief description of
the relevant proofs. The chronological section treated the Hijra, Yazdegird,
Seleucid, Maliki (or Jalali), Chinese-Uighur, and Ilkhani eras and calendars.
The trigonometric tables were calculated to four places for each minute of arc.
For spherical trigonometric functions the tables were computed to two places
for each degree of solar mean longitude. Mean motions of the planets were
provided to thirds of an arc for Yazdegird years 781 through 791. Motions were
also given for Yazdegird years 10, 20, 30 . . . 100, and appropriate numbers
were supplied for months, days, and hours. Processional motions were tabulated
for months and to three fractional places.

There was a complete set of tables for planetary equations and latitudes and
an extensive table (Kashi's contribution) for simplifying the computations. The
zij contained tables showing the maximum and minimum duration for each
planet of forward and retrograde motions, expressed in days (carried to hours)
and degrees (carried to seconds). For planetary sectors there were tables for epicycles and deferents computed to minutes of arc for both velocity and distance.
A table for adjusted lunar parallax calculated to two places was provided along

with an interpolation scheme for modifying the values of the lunar anomaly. Eclipse tables in both the ecliptic and the plane of the orbit were given to two places for each five degrees. There was also a table of mean conjunctions and oppositions for the Hijra years 801, 802, ... 811 and for 10, 20, 30, ... 100, 200, 300 Hijra years. Entries were calculated to four fractional places of time and to seconds of arc for the conjunction or opposition. For visibility there were tables of lunar settings computed to seconds of arc for each zodiacal sign, and for the planets there were tables for the third and fourth climes only. The latitude and longitude of 516 localities (divided according to the seven climes) were given, the coordinates terminating in multiples of five minutes. For eighty-four stars the latitude, longitude, magnitude, and temperament were provided. The coordinates, taken either from the *Almagest* or the *Zij-i Ilkhani*, were corrected for precession. Al-Kashi also included a great deal of astrological material: on nativities, anniversaries, and monthly and annual progressions of the zodiacal signs. He prepared tables for the life periods of the planets and for diurnal and nocturnal nativities. There was also a table of world years for 301, 302, 303 . . . 402 of the Maliki (or Jalali) era and for every 100, 200, 300 . . . 1000, 2000, 3000 . . . 10,000, 20,000, 30,000 . . . years. A final table listed the small, middle, large, and mighty periods for the world year cycle. The *zij* concluded with a table listing the distances of the planets from the centre of the Earth. Additional tables showed the interpolations required for the various deferent and epicycle positions.[18]

Although in Samarqand al-Kashi's most important scientific contributions were theoretical, his first major assignment seems to have been practical – overseeing the planning, construction, and equipping of the new observatory. In 1416, just before arriving in Samarqand, he had written two treatises on astronomical instruments: *Treatise on Observational Instruments* and *The Garden Excursions*. Nevertheless, Ulugh Beg had to be consulted. Al-Kashi wrote:

> The developments to date in the activity pertaining to the observatory have been in accordance with what this servant proposed to His Majesty . . . in the case of the observatory building and . . . each instrument. His Majesty reflected upon these recommendations . . . Whatever he approved he ordered to be carried out, and as to other cases, he enriched them with new ideas and inferences and ordered the adoption of the modified versions . . .[19]

The observatory was constructed on a hill twenty-one metres high outside the city. Its top surface measured 170 metres (north to south) by 85 metres (east to west). The main building, which housed the giant meridian arc (or mural sextant), was cylindrical. Built of solid brick and limestone, its exterior was faced with the same coloured tile that had been used in the madrasa. It was set in a large garden and was surrounded by a collection of small dwellings for students and staff.[20] Ulugh Beg's meridian arc was modelled after the Suds-i Fakhri

(Fakhri Sextant), the famous mural sextant built by Abu Mahmud al-Khujandi in 994 for the Buwayhid ruler Fakhr al-Dawla. At Samarqand a trench two metres wide was dug in the hill, and in it an arc was placed. The sextant was eleven metres long and fully calibrated. The arc had a radius of about 40 metres, 60 per cent above ground and the rest below. The instrument was used to observe the Sun, Moon, planets, and stars. With it Ulugh Beg's astronomers were able to more accurately determine the obliquity of the ecliptic. Their value – 23.52 degrees – was more accurate than those of Copernicus or Tycho Brahe. When finally completed, the *Zij al-Sultani* boasted the most accurate astronomical and astrological tables in the world.[21]

The giant meridian arc was excavated in the early twentieth century and has been photographed and thoroughly studied. Of the other instruments approved by Ulugh Beg and erected by his builders, less information is available. In his *Treatise on Observational Instruments* al-Kashi described the parallactic ruler, the armillary sphere, the equinoctial armilla, the instrument with two rings, the Suds-i Fakhri, the azimuthal quadrant, and the sine and versed sine instrument. And, in a letter to his father, he boasted that the observatory contained an azimuthal quadrant, two armillary spheres (one fabricated by a coppersmith named Ibrahim), an instrument with two holes, and a large astrolabe.[22] In the sixteenth-century one historian described a sine and versed sine instrument at Samarqand,[23] and in the eighteenth-century Maharajah Jai Singh, the builder of the Shahjanabad Observatory in India, wrote of 'instruments such as those which were constructed in Samarqand . . . armillary sphere, instrument with two holes, parallactic ruler, the suds-i Fakhri, and the al ala al shamila.'[24] A later historian described beautiful models of the celestial spheres – depicting the location of the stars and the epicycles of the planets. There was also a terrestrial globe, showing mountains, seas, deserts, and the seven climes.[25] Al-Kashi wrote that the water clocks in the observatory consisted of bowls with holes in the bottom. They floated on the water surface and became gradually filled, sinking slowly with the passage of time.[26]

The observatory also contained a library. Once, confused by a proof from al-Biruni's *Masudic Canon*, Qadi Zada picked the book from the library shelf and took it back to his room to study. At another point, unsure about the correct formula to apply for a certain calculation, he consulted a textbook from the library, following step-by-step the directions set out. And, according to its inscription, *Key to Arithmetic*, one of al-Kashi's most popular books, was written for Ulugh Beg's library.[27]

Observational work began as soon as possible after ground was broken. In Chapter Three of the *Zij al-Sultani* Ulugh Beg described the programme he had decided upon: the Sun and Moon were to be observed daily, Mercury every fifth day, and the rest of the planets every tenth day. During periods of planetary reversal observations were to be made daily.[28] Transforming these raw numbers into astronomical and astrological tables was not easy; it required a sophisticated knowledge of complicated formulas and equations. Given the number and size of the instruments and the difficulties of calculation, a large number of mathematicians and astronomers (as many as sixty or seventy) were required for the day-to-day work of observation, measurement, and calculation. Of the Samarqand staff al-Kashi wrote:

> Although there are many people here who are conversant with the mathematical sciences, none of them is such that he is acquainted with both the theoretical ('scientific') and the applied ('practical') sides of observations . . . Applied astronomy . . . is divided into scientific and practical. The practical side of applied astronomy may be illustrated with the following example. Suppose that two stars have reached the first perpendicular at a certain condition. Elevation is measured with an instrument, and the latitude and longitude of one of these stars is known. It is required to derive the latitude and longitude of the other star from these data . . . The knowledge of how to derive this, i.e., to know to multiply which quantity with which and to divide by what and how to proceed in order to obtain the desired result constitutes the scientific side of this operation (of applied astronomy). The scientific side of theoretical astronomy ('the absolutely scientific') is the knowledge of the science itself . . .[29]

Al-Kashi was the acknowledged leader of the observatory in its first decade and seems to have been one of the few men who had mastered both the theoretical and the practical aspects of astronomy. He had already compiled a *zij* (given to Ulugh Beg in 1416), and he was skilled in the design, construction, and operation of observational equipment. Al-Kashi, however, died (1429) a decade before the *Zij al-Sultani* was finished and two decades before the death of Ulugh Beg, but in his eight or nine years at the observatory he was busy and productive. After the meridian arc was installed, he took a regular turn at the sextant, making observations and recording results.

More important, however, than the applied work were his theoretical writings. For it was after the new observatory had become fully operational that he composed the three most important treatises of his long and illustrious career. In 1424 he finished *Treatise on the Circumference*. In the preface he pointed out that the values for π (the ratio of the circumference of a circle to its diameter) obtained by his predecessors had been inexact, resulting in substantial errors when calculating the circumference and area of large circles. With the giant meridian arc he was able to make more precise measurements and to calculate the value of 2π more accurately – to sixteen decimal places. Al-Kashi's result was a major advance: Archimedes' fraction had been 3.14, Ptolemy's 3.14166, and a Chinese scholar had achieved six places. In 1597, van Roomen calculated π to fifteen decimal places but it was not until 1615, nearly 200 years later, however, that the Dutchman L. van Ceulen would surpass al-Kashi by calculating the fraction to twenty places.[30]

Al-Kashi's second and best-known treatise – *Key to Arithmetic* – was finished in Samarqand in 1427, two years before his death. It was an encyclopaedia of elementary mathematics and was intended for a wide range of students – astronomers, land surveyors, architects, accountants, clerks, and merchants. In its breadth and elegance, al-Kashi's *Key* won a wide audience; extracts and compendia were produced for centuries. He defined arithmetic as the 'science of finding numerical unknowns with the aid of corresponding known quantities'. The work was divided into five books: 'Arithmetic of Integers', 'Arithmetic of Fractions', 'On the Computation of Astronomers' (on sexagesimal arthimetic), 'On the Measurement of Plane Figures and Bodies', and 'On the Solution of Problems by Means of Algebra'. Especially noteworthy in the second and third books was al-Kashi's account of decimal fractions. Although earlier mathematicians had touched on decimal fractions, al-Kashi was the first to work out a methodical introduction to the topic. His aim was to establish a system of fractions in which operations could be carried out in the same manner as with integers, and he worked out the analogies between both systems of fractions – the decimal and sexagesimal. At another point in this treatise, al-Kashi developed an algorithm for calculating roots to the nth power.

Al-Kashi's third and last treatise, finished just before his death in 1429, was *Treatise on the Chord and Sine*. With the Samarqand instruments he was able to calculate the value of sine 1 degree to ten sexagesimal places. The best previous result had been correct to only four places.[31]

Jamshid al-Kashi died in 1429. And, although he had been a productive member of the observatory staff during its first decade, making good use of the accuracy provided by the new equipment, he was not a major participant in the observational programme that resulted in the *Zij al-Sultani* (c. 1440). In his preface Ulugh Beg wrote, 'The work was started jointly with the aid of Qadizada-i Rumi . . . and Giyath al Din Jamshid . . . At the initial stage of the work . . . Giyath al Din Jamshid . . . passed away . . . Thereafter the work was completed by 'Ali ibn Muhammad Qushji.'[32]

After the death of al-Kashi, Ulugh Beg put Qadi Zada and 'Ala al-Din 'Ali ibn Muhammad al-Qushji (1403–74) in direct charge of the observatory and its staff. Qadi Zada, as we have seen, was primarily a teacher and, while knowledgeable enough, did not have the creative spark of either al-Kashi or al-Qushji. During the observatory's second decade Qadi Zada continued to head the madrasa. At the observatory he was involved in the creation of a new star catalogue, an update of the one in Tusi's *Zij-i Ilkhani*. It contained new and more accurate descriptions of some 1,018 stars – their positions, magnitudes, and colours. Qadi Zada, however, died c. 1440, just as the finishing touches were being put on the first version of Ulugh Beg's treatise.

Of the four principal astronomers of the Samarqand Observatory (the above three and Ulugh Beg), 'Ali al-Qushji was the youngest. The son of Ulugh Beg's falconer (or *qushji*), he was extraordinarily gifted. His early education was at the Samarqand Madrasa, where he attended the lectures of Qadi Zada, al-Kashi, and eventually Ulugh Beg himself. In 1420, just turned seventeen, he moved to Kirman to study mathematics and astronomy with Mulla Jami. As a result, he was absent during the construction and initial years of the new observatory. In 1428, however, he returned to Samarqand and joined the staff of the new institution, presenting Ulugh Beg with an impressive treatise on the motions of Mercury that corrected the Ptolemaic theory. Al-Qushji remained in Samarqand for the next twenty years, overseeing the work of the observatory, attending to the publication and later additions to the *Zij al-Sultani*, and composing his own works. After Ulugh Beg's murder in 1449, he moved to Herat, where he dedicated a commentary on Tusi's *Summation of Belief* to Sultan Abu Sa'id Mirza (1449–69). In 1469 he relocated to Tabriz and the court of the Aq Qoyunlu ruler Uzun Hasan (1467–78), and in 1472 he moved again to Istanbul and the court of Mehmed II (1441–6, 1467–81). In the remaining three years of his life he taught first at the Sahn-i Tahman Madrasa (founded by Mehmed II) and then was appointed head of the Ayasofya Madrasa.[33]

'Ali al-Qushji had a long and illustrious career – twenty years at the Samarqand Observatory with Ulugh Beg and twenty-four more years in Herat, Tabriz, and Istanbul. Over the course of his career he composed nine works on astronomy – original contributions along with textbooks and summaries. His two textbooks – *Treatise on the Science of Astronomy* and *Treatise on Arithmetic* – were very popular. His most important contribution, however, was his work on the *Zij al-Sultani*. After the deaths of both al-Kashi and Qadi Zada, Ulugh Beg put al-Qushji in charge of the day-to-day activity of the observatory – the manning of instruments, the recording of results, and the preparation of the various tables. The production of a completely new *zij* was, as we have seen, a lengthy affair. Because it took Saturn thirty years to complete its circuit of the Sun, an entirely new set of tables would have to include an observational programme of at least that length.

For both 'Umar Khayyam at Isfahan and Nasir al-Din Tusi at Maragha a thirty-year programme of observation and calculation had proven impossible. Malik Shah had been unwilling to fund such a lengthy affair and, while Hulagu and his successors kept Maragha operational for more than fifty years, the *Zij-i Ilkhani* was finished two years before Tusi's death, after no more than thirteen years of work in the observatory. For Ulugh Beg, on the other hand, as much astronomer as ruler, the completion of a new treatise was a career achievement. He himself had been intimately involved in the twenty-year period (1420–40) that preceded the publication of the *Zij al-Sultani*, and he lived for another ten years, during which a number of additions and corrections were added. For the final twenty years of the prince's life, al-Qushji supervised the activities of the observatory. He was deeply involved in the initial draft of the treatise, and he supplied (in his later *Shahr-i Zij al-Sultani or Commentary on the Zij al-Sultani*) the final ten years of corrections and additions, transforming the Samarqand *zij* into the most complete and accurate in the world.[34]

The superiority of the *Zij al-Sultani* was due primarily to two things. Firstly, the theoretical contributions of al-Kashi – the use of decimal fractions and the more precise calculations of the values of sine 1 degree and π. Secondly, the new and more accurate observations of the planets and stars made possible by the outsized and sophisticated equipment of the observatory.[35] The treatise itself was divided into the usual sections. The chronological tables covered the Hijra, Yazdegird, Seleucid, Maliki (or Jalali), and Chinese-Uighur eras and calendars. The trigonometric tables were calculated to five places for both the sine and tan functions and the spherical trigonometric functions were computed to three places. The tables for the equations of time were expressed to three fractional places. The mean motions of the planets were shown for Hijra years 841, 842, 843 . . . 871 and also for 30, 60, 90 . . . 300, 600 900 . . . 1200 Hijra years and for the appropriate months, days, hours, and minutes. There was also a table for determining the effects of change in terrestrial longitude – all expressed in seconds of arc. Planetary equations and latitudes, including a table of the third lunar equation, were computed to thirds of an arc. The tables for the stations and retro-gradations of the planets were identical to those in al-Kashi's *zij*, and the planetary sectors were computed to seconds of arc. For planetary sectors there were tables for all four categories computed to seconds of arc. The tables for the solar and lunar parallax in the altitude circle were computed to three places. An interpolation scheme allowed a modification of the values for the variation in the lunar anomaly. Tables for both solar and lunar eclipses included magnitude and duration of immersion and totality and were tabulated to two places. The lunar visibility tables, embracing the lunar equation for setting, were computed to minutes of arc for each zodiacal sign. The planetary visibility tables were the same as al-Kashi's. The latitudes and longitudes of 240 towns and cities, arranged by regions rather than by climes, were tabulated to minutes of arc. A new star table, the result of mostly fresh observations, gave ecliptic coordinates for 1,018 stars. An astrological section dedicated to nativities and world periods repeated the material in al-Kashi's *zij*. A final section contained a table of the greatest and least lunar distances with an interpolation scheme computed to three places. There was also a table of solar distances calculated to three places.[36]

The Samarqand Observatory was the largest and most sophisticated in the Eurasian world, and the *Zij al-Sultani* contained the most complete and accurate set of astronomical tables. And the primary reason for these achievements was the founder of the observatory and the author of the *zij*, Ulugh Beg himself. Although in the Islamic world a major astronomical programme depended on the interest and support of a ruler (al-Ma'mun, Malik Shah, Hulagu, or Alfonso X, for example), Ulugh Beg was different: his passion was deeper and his training and commitment were greater. The Timurid prince's desire for a more accurate set of astronomical tables was not primarily astrological, rather he was a legitimate scientist with a formidable intellect and a rigorous training in mathematics and astronomy. Al-Kashi praised his memory and arithmetical skills and his expertise with instruments, observations, and calculations. A minor but telling example of Ulugh Beg's talent was found in a marginal note scribbled on the observatory's copy of al-Kashi's *Zij-i Khaqani*. After praising extravagantly the author of the treatise – Ptolemy would have been incapable of comprehending al-Kashi's planetary configurations and Abu Ma'shar would have been equally inept with his theorems – Ulugh Beg pencilled in an ingenious new technique for calculating the distance between two stars given their ecliptic coordinates.[37] The other reason for the success of the Samarqand Observatory, in addition to Ulugh Beg's scientific capability, was his skill as an administrator. For nearly forty years he recruited the finest scientists and artisans of West and Central Asia. The Samarqand Observatory provided the best training, the newest observational instruments, and the most comfortable and collegiate atmosphere.

As a result, the observatory was enormously productive. It was the only institution whose founder directed and supported observational work for the thirty years that were required for a complete set of fresh observations. Tusi's observatory in Maragha had remained operational longer than Ulugh Beg's but it went through a succession of rulers, astronomers, and supervisors. In Samarqand, on the other hand, not only was Ulugh Beg in charge of the Madrasa and the Observatory for nearly thirty years, but the work of al-Qushji in the final twenty years – revising, supplementing, and correcting – meant that the *Zij al-Sultani* was the first astronomical treatise since the *Almagest* that was based on an observational programme of the magical thirty years.

In observational astronomy, the Samarqand Observatory was also at the forefront. Its most conspicuous feature was its giant meridian sextant. The bigger the distance between successive degrees on the scale, the more graduations, and the greater the precision. Although the advantage of size was sometimes offset by construction errors, Ulugh Beg's instrument fulfilled its promise. Its value for the latitude of Samarqand was extremely accurate: 39 degrees, 37 minutes, and 23 seconds – within two minutes and five seconds of the modern value. And his determination of the obliquity of the ecliptic was the most accurate yet. Finally, the large meridian sextant enabled the Samarqand astronomers to produce an entirely new star catalogue. The record in the *Zij al-Sultani* contained coordinates for 1,018 fixed stars and was the only catalogue between Ptolemy's in the second century and Tyco Brahe's in the sixteenth to be based on original observations.[38]

In theoretical astronomy also, the Samarqand astronomers were ahead of their time. Building on the contributions of Nasir al-Din Tusi and his collaborators, Ulugh Beg's men laboured over the problem of constructing planetary models that would give accurate numerical results while adhering to the principle of uniform circular motion – via modifications such as the 'Tusi Couple'.[39]

Finally, computational mathematics was an important focus for Ulugh Beg and his collaborators. Al-Kashi increased the precision and quantity of numerical resources available for astronomers. Both he and Qadi Zada provided algorithms for calculating the sine of one degree to a very high degree of accuracy. And in the sine table of the *Zij al-Sultani* there was an entry for each minute of every degree – a total of 10,800 entries, each entry carried to five sexagesimal places. Al-Kashi also introduced decimal fractions. He was the first to work out, explain, and apply a system of decimal fractions (in addition to sexagesimal) to the entire field of astronomical mathematics. According to a prominent scholar 'Ulugh Beg's observatory was carrying out the most advanced observations and analysis being done anywhere. In the 1420s and 1430s Samarkand was the astronomical and mathematical capital of the world.'[40]

Notes

1. Sayili, *Observatory*, 228–32.
2. Ibid. 236–42.
3. Hockey, 'Ibn Shatir', in *Biographical Encyclopedia of Astronomers*, 569–70.
4. Sayili, *Observatory*, 245.
5. Saliba, *History of Arabic Astronomy*, 233–40; Sayili, *Observatory*, 245–6; North, *Cosmos*, 208.
6. Kennedy, *Islamic Astronomical Tables*, 125, 163–4.
7. Hockey, 'Ulugh Beg: Muhammad Taraghay ibn Shahrukh ibn Timur', *Biographical Encyclopedia of Astronomers*, 1157–9.
8. Sayili, *Observatory*, 262.
9. Ibid. 268.
10. Ibid. 262.
11. Ibid. 262, 280–1.
12. Ibid. 262–3, 266–8, 272–3.
13. Ekmeleddin Ihsanoglu (ed.), *History of the Ottoman State, Society, and Civilisation*, 2 vols (Istanbul: Research Center for Islamic, Art, and Culture, 2001–2), 2: 521–23; Hockey, 'Qadizade al Rumi: Salah al Din Musa ibn Muhammad al Rumi', in *Biographical Encyclopedia of Astronomers*, 942.
14. Sayili, *Observatory*, 262.
15. Ibid. 224–5.
16. Ibid. 250–2.
17. 'al-Kashi (or al-Kashani), Ghiyath Al Din Jamshid Masud', *Dictionary of Scientific Biography*.
18. Kennedy, *Islamic Astronomical Tables*, 128, 166.
19. Sayili, *Observatory*, 263.
20. Ibid. 274–5.
21. Ibid. 283.
22. Ibid. 282, 286.
23. Ibid. 287.

24. Ibid. 283.
25. Ibid. 282.
26. Ibid. 288.
27. Ibid. 281.
28. Ibid. 280.
29. Ibid. 250.
30. 'al-Kashi', *Dictionary of Scientific Biography*.
31. Ibid.
32. Sayili, *Observatory*, 266.
33. Hockey, 'Ihsan Fazlioglu, Qushji, Abu al Qasim 'Ala al-Din Muhammad Qushci-zade', in *Biographical Encyclopedia of Astronomers*, 946–8.
34. Ibid.
35. North, *Cosmos*, 211.
36. Kennedy, *Islamic Astronomical Tables*, 166–7; see also Saliba, 'The Astronomer al-Sufi as a source for Ulug Beg's Star Catgalogue (1437)', in *A History of Arabic Astronomy*, no. XII.
37. Edward S. Kennedy, 'Ulugh Beg as Scientist', in Kennedy, *Astronomy and Astrology in the Medieval Islamic World*, 1–3.
38. Kennedy, 'Heritage of Ulugh Beg', in Ibid. 4–5.
39. Ibid. 7.
40. Ibid. 11.

CHAPTER SEVEN

The observatory in Istanbul

The Samarqand Observatory (begun in 1420) and the *Zīj al-Sultani* (completed in 1440) were the high points in the history of Islamic astronomy. After Samarqand observatories were founded in the two great early-modern Islamic empires – one in the Ottoman Empire in 1577 and five in the Mughal Empire between 1728 and 1734. In neither state, however, did the Ottoman or Mughal astronomers substantially surpass the achievements of Ulugh Beg and his men – not in instrument design, observational accuracy, or mathematical creativity. Nevertheless, the Ottoman and Mughal astronomers played an important role in the development of the heavenly sciences as they grappled with the flawed system they had inherited, working to bridge the gap between the Ptolemaic, geocentric theories of the first millennium and the Copernican, heliocentric concepts of the late second millennium.

The Ottoman Empire (c. 1300–1923) was the first and longest lived of the three early-modern states that ruled west and south Asia from c. 1500 to 1800. Unlike the other two – the Safavid Empire (1501–1722) in Iran or the Mughal Empire (1526–1739) in India – the Ottomans did not rule a defined geographical entity. Rather, by the late sixteenth century their empire included an enormous but heterogeneous collection of lands and people in the eastern Mediterranean, stretching from Iraq and Anatolia in the east through Syria, Egypt, the Arabian peninsula, Hungary, and the Balkans to North Africa in the west. Osman (1281–1326), the founder of the dynasty, was the ruler of a small Seljuq successor state in western Anatolia. He and his followers were Muslim Turks, descendants of the Central Asian Seljuqs who had defeated the Abbasids at Baghdad (1055) and had established a successful and far-flung dynasty. Malik Shah, builder of the Isfahan Observatory, had been one of the most important Seljuq rulers. In the early fourteenth century Osman and his warriors expanded their nascent state east across Anatolia and west into the Balkans. Under Bayezid I (1389–1402), however, their advance was halted. Timur, the great Central Asia conqueror, defeated the Ottomans at Ankara (1402), capturing and later killing Bayezid.[1]

The first half of the fifteenth century, between the death of Bayezid I (1402) and the accession of Mehmed II (1444–6, 1451–88), was a time of disruption and dissension. Timur had ended Ottoman control of Anatolia, and it was not easily re-established. Although Mehmed II conquered Constantinople in 1453 and expanded the boundaries of the state in both the East and the West, it was Selim I (1512–20) who captured Syria, Egypt, and the Arabian peninsula, transforming the fledgling Ottoman state into the leading Islamic empire in the world. Selim's son, Suleiman I (1520–66) or Suleiman the Magnificent (as he was known in Europe), propelled the empire to the peak of its military, political, and economic power. He extended Ottoman rule into Eastern Europe, conquering Hungary and just failing in 1529 to capture Vienna. Suleiman's son Selim II (1566–74) and grandson Murad III (1574–95) followed him to the throne.

After his capture in 1453 of Constantinople (later Istanbul), Mehmed II reorganised the Ottoman system of higher education – establishing a new group of madrasas that taught the rational sciences – mathematics, medicine, logic, philosophy, and astronomy – as well as the religious. Al-Qushji, the Samarqand astronomer, who came to Istanbul during the last years of Mehmed's life, probably influenced the ruler's decision. On the rational curriculum a contemporary observed:

> As . . . Geometry and arithmetic are easily apprehendible subjects and because they do not require much deep thought [they] are not studied as separate subjects . . . Because astronomy involves the use of the imaginative powers and supposition and is therefore more difficult than geometry, they study it later as a separate subject . . . Even there they do no remain idle but undertake discussions of arithmetic, geometry, astrolabes, rub (quarter), land surveying . . . [2]

Despite the new madrasas, from the late fifteenth-century onward the significant work in Ottoman astronomy was done in one of three other places: in the office of the chief imperial astronomer, in the office of the local timekeeper, or in the Istanbul Observatory.

The office of the chief imperial astronomer was established by Bayezid II (1481–1512).[3] As a palace official, the chief astronomer was a member of the learned class, having been trained in one of Mehmed II's new madrasas. His office included other astronomers as well – a group of second level officials along with four or five highly trained assistants. The principal duty of the chief imperial astronomer was to prepare the annual almanac (*takwim*) for the sultan and his court. Like the astronomer/astrologers of the Abbasid, Seljuk, Ilkhanid, and Timurid courts, he advised the sultan on the proper times for inaugurations, weddings, circumcisions, fasts, and journeys. He was also called upon to predict and interpret heavenly events – comets, eclipses, and conjunctions.[4]

The timekeeper (*muwakkit*) of the local mosque, in the Ottoman Empire as in the early Islamic states, was responsible for determining the times of the five daily prayers. During the eighth century these times had become standardised: the muezzin sounded his call at sunset, evening (after twilight), morning (at daybreak), noon, and afternoon (when the Sun was high). Over the centuries timekeeping technology became more sophisticated and accurate. The gnomon was replaced by the sundial, which gave way to the clepsydras (water clock) and finally to the mechanical clock. The first Ottoman timekeepers were found in Istanbul, attached to the imperial mosques constructed by Mehmed II and his successors. Since timekeepers were normally astronomers and knew how to use simple astronomical instruments, their houses often became small observatories and, when enough interest was shown, temporary classrooms. Although their salaries and expenses were met from the endowment (waqf) funds of the local mosque, it was the chief imperial astronomer who was responsible for their appointment and regulation. A son could follow his father but if none was available, an exam would determine the most qualified candidate. In fact, timekeeping was often the first step on the ladder of advancement for the newly-minted astronomer: small mosque to larger mosque, to the staff of the chief imperial astronomer and if, talented enough, to the top spot itself.[5]

The third astronomical institution in the early-modern Ottoman Empire was the Istanbul Observatory. Founded in 1575 by Taqi al-Din, chief astronomer of Murad III (1574–95), it was the last Islamic observatory capable of cutting-edge research. And its director was the last in a long line of scientific polymaths – a lineage of astronomical and mathematical geniuses that stretched from 'Umar Khayyam and Nasir al-Din Tusi to Jamshid al-Kashi and Taqi al-Din himself. Before turning to Murad III's chief astronomer, however, it is necessary to look at two transitional figures, Qadi Zada and al-Qushji. These two early Ottoman astronomers, whom we have already met, exemplified the scientific and personal relationships that connected Samarqand, on the one hand, with Istanbul on the other.

Although Qadi Zada was born, raised, and educated in Ottoman Bursa, the bulk of his scientific life was spent in Samarqand, working in the madrasa and observatory founded by Ulugh Beg. And his writing was almost entirely in Arabic – a few works in Persian, none at all in Ottoman Turkish. Al-Qushji, the most important mathematician/astronomer under Mehmed II, was also an outlier. He was neither Ottoman born nor, until the last two years of his life, was he an Ottoman subject. And, like Qadi Zada, he spent a large part of his scientific life in Samarqand – the early years of his education at the madrasa and twenty later years as co-director of the observatory.

Al-Qushji, however, did have an impact on Ottoman science. In 1469 he was sent by the Aq Qoyunlu ruler, Uzun Hasan, on a diplomatic mission to Istanbul. Al-Qushji so impressed Mehmed II, who was in the middle of his educational reform, that he was offered a lavish salary and a prestigious post in the Ottoman capital. On his arrival in 1472 Mehmed II put him in charge of the Ayasofya Madrasa, and although he lived only three more years, al-Qushji managed to compose two more treatises. The first, *Muhammad's Treatise*, was written on his way from Tabriz to Istanbul and was dedicated to his patron-to-be. It was an expanded Arabic version (six chapters instead of three) of his earlier Persian *Treatise on the Science of Arithmetic*. Al-Qushji's second work was *The Victory Treatise*. Composed in 1473 on the occasion of the Ottoman victory over Uzun Hasan, this was a longer Arabic reworking of his 1458 Persian work, *Treatise on the Science of Astronomy*. The new version contained three chapters: the first on the planets (number and locations), the second on the Earth (its shape and the various climes), and the third on the clouds and other heavenly bodies. It also became a standard madrasa textbook and attracted many commentaries, including one by al-Qushji's grandson, Miriam Celebi (d. 1524–5). In 1579–80 it was translated into Ottoman Turkish.[6]

The greatest Ottoman astronomer, however, was Taqi al-Din ibn Ma'ruf (1526–85). Born in Damascus, his early education was in the traditional sciences. But, like many astronomers before him, he was soon attracted to the rational sciences. His father, Ma'ruf Efendi, encouraged his scientific interests, and he studied with two famous scholars – in Damascus with the mathematician Shihab al-Din Ghazzi, and in Cairo with the astronomer Muhammad ibn Abi al-Fath al-Sufi (fl. late fifteenth and early sixteenth century). Al-Sufi was particularly significant because he had drafted a commentary on Ulugh Beg's *Zij al-Sultani*, recalculating the tables for Cairo. After completing his education, Taqi al-Din taught for several years in the madrasas of Damascus and Cairo. The year 1550, however, marked an important turn. In the early months he and his father travelled to Istanbul, meeting some of the leading Ottoman scientists. Returning to Cairo, Taqi al-Din taught for several months. But Istanbul called, and in the Ottoman capital he caught the eye of Grand Vizier Samiz 'Ali Pasha and was appointed head professor of the chief madrasa in Edirne. When 'Ali Pasha was appointed governor of Egypt, Taqi al-Din followed his patron back to Cairo. Interested in astronomy and mathematics himself, 'Ali Pasha recognised his protégé's talent and gave Taqi al-Din unfettered access to his library and his extensive collection of mechanical clocks. In addition, Qutb al-Din Zada Mahmud, grandson of al-Qushji, also offered Taqi al-Din his library – works by al-Qushji, Jamshid al-Kashi, and Qadi Zada – and his collection of astronomical instruments. Although he worked as a judge and teacher in the Ottoman province of Egypt for the next twenty years, Taqi al-Din found time to write and soon established a solid scientific reputation. He returned to Istanbul in 1570 and in 1571, on the death of Mustafa ibn 'Ali al-Muwaqqit, was appointed chief astronomer by Selim II (1566–74).

Like the other polymaths, Taqi al-Din was enormously productive. The (incomplete) record of his oeuvre totals forty-six: six in mathematics, thirty-three in astronomy, four in mechanics, and three in optics. Only a few of his treatises are dated or datable but a rough division would place the ones on astronomy, mathematics, and optics in the Istanbul period and the ones on mechanics and clocks in the Egyptian.

Soon after returning to Egypt in 1550, Taqi al-Din was sent as a judge to Nablus in Ottoman Palestine. There he completed a treatise on mechanics: *The Sublime Methods in Spiritual Devices.* It contained six chapters: the first, on the clepsydras (water clock); the second, on the rotating spit; the third, on fountains; the fourth, on methods of irrigation; the fifth, on devices for lifting weights; and the last, on devices for raising water. In the first chapter he pointed out the drawbacks of water clocks and commented briefly on an Arabic work on mechanical clocks – both of these a preparation for his own groundbreaking treatise on mechanical clocks. To power his rotating spit, Taqi al-Din designed an early version of the steam turbine. And to lift water from one irrigation channel to another, he drew up plans for a six-cylinder pump. A sophisticated device for its time, the pump was intended to replace the traditional animal-powered Persian wheel.[7]

The most important of Taqi al-Din's pre-Istanbul treatises, however, was *The Brightest Stars for the Construction of Mechanical Clocks.* Finished in 1559, it was dedicated to 'Ali Pasha. Like the Nablus works on the steam turbine and the water pump, this one was marked by technical ingenuity and careful design. Unlike them, however, Taqi al-Din's mechanical clocks actually completed the transition from drawing board to real world. Mechanical clocks had first appeared in Europe in the early sixteenth century (the first spring-powered pocket watch in 1524), and European domestic clocks had become fairly common in Istanbul palaces and mansions by mid-century. Taqi al-Din divided his treatise into five chapters: the tower clock, the pocket clock, the domestic clock, the astronomical clock, and the observational clock. Given his interests, it is not surprising to find that he devoted most of his discussion to the last two varieties.[8]

For Taqi al-Din, the principal drawback of the traditional timekeeping devices (the sundial and the water clock) was their inaccuracy. But after testing the European imports in 'Ali Pasha's collection he found them lacking also – losing as much as twenty minutes a day or more. In Taqi al-Din's innovative design, the clocks included one or more escapements (a cogged wheel that transferred energy to the timekeeping mechanism), several alarms, and a power system of falling weights. His astronomical clock displayed the phases of the Moon, the days of the week, the relationships between the Sun and Moon, the position of the Sun in the zodiac, the azimuths, latitudes, and ascensions of certain stars, and the times of prayer. And his observational clock featured three dials – for hours, minutes, and seconds. In their day, Taqi al-Din's clocks were the most accurate and sophisticated in the Eurasian world, allowing Ottoman astronomers to time and chart the movements of the heavenly bodies with unparalleled precision.[9]

Taqi al-Din came to Istanbul in 1570 and was appointed chief imperial astronomer one year later. Although the bulk of his astronomical and mathematical work was done after his arrival in the Ottoman capital, he must already have made a name for himself – in order to have achieved so rapid and important a promotion. In Cairo and Ottoman Palestine he had recorded a number of celestial observations (employing an astronomical well 25 metres deep) and drafted a treatise on sundials.[10] In the four years before the accession of Murad III in late December 1574, Taqi al-Din was extremely busy. With a large staff of at least fifteen astronomers and a sizable collection of portable instruments, he began a detailed record of celestial phenomena. Although the exact location of these observations is unknown, Taqi al-Din and his men likely made use of the giant Galata Tower. Erected in 1348 in the Genoese quarter of Constantinople, it was 220 feet tall.

On the decision to found an observatory in Istanbul the sources are not very clear: some attribute the initiative to Murad III and some to Taqi al-Din himself. However, a careful reading of the evidence, especially of 'Ala al-Din al-Mansur's *Shahinshahnama*, a Persian chronicle in verse of the first years of Murad's reign, suggests a more complicated mix of motives. 'Ala al-Din was from Shiraz, the famous city of poets, and finished his work in October 1581. According to him, at least four persons were involved. In the first place, Taqi al-Din, appointed chief imperial astronomer in 1571, nearly four years before Murad's accession on 21 December 1574, had a large staff and had already conducted a series of celestial observations. And he seems to have harboured a strong desire to construct a well-equipped observatory that would allow him to update the *Zij al-Sultani* of Ulugh Beg – more than 130 years old at this point. Al-Mansur wrote:

> The determination of planetary positions, as well as the construction of
>   astronomical tables,
> Is made by those well-versed in that science.
> It is through this science that one comes to know with certainty
> The mansions of the sun and the moon in the signs of
> Tauris and Aries
> And if the moon is found to be in the sign of Scorpio,
> Marriage must certainly be avoided at such a time[11]

In addition the earlier *zij*s were out-of-date:

> The astronomical tables of Ulugh Bey [*Zij al-Sultani*] and Nasir al Din Tusi
>   [*Zij-i Ilkhani*]
> Had become worn-out like traces of mats upon soft soil[12]

The second important person in the story of the Istanbul Observatory – in its founding and construction as well as in its operation and eventual destruction – was the all-powerful Grand Vizier Sokollu Mehmed Pasha. Like many other high-ranking Ottoman officials, he had been part of the annual levy of Christian boys. Born in Bosnia to an Eastern Orthodox family, he was converted to Islam and educated in the imperial school at the Topkapı Palace in Istanbul. Among these young servitors, a true meritocracy reigned, and Sokollu Mehmed's talents propelled him rapidly upward. Beginning as an apprentice member of the Janissaries (private soldiers of the sultan), he was at 29 promoted to the Imperial Treasury. At 35 he became Imperial Chamberlain and at 37 commander of the imperial guard. He soon came to the attention of the reigning sultan, Suleiman the Magnificent (1520–66), and in 1546 was appointed High Admiral of the Fleet. In 1551 he became governor-general of Rumelia (Balkan provinces of the Ottoman Empire), in 1555 Third Vizier, and in 1561 Second Vizier. In 1562 he married Suleiman's granddaughter, and in 1565, a year before the Sultan's death, he was elevated to the post of Grand Vizier. From 1566 until his death in 1579 Sokollu Mehmed Pasha was the most powerful man in the empire. As a result, he was the first person Taqi al-Din approached. A general and statesman rather than a scientist, the Grand Vizier gave a general assent to the proposal and directed the chief astronomer to Khwaja Sa'adat al-Din (d. 1599), Murad's one-time tutor.[13]

In contrast to the vizier, Khwaja Sa'adat was a distinguished intellect, having mastered both the traditional and the rational sciences, and many of his students had achieved renown in mathematics and astronomy. Of him 'Ala al-Din al-Mansur wrote (with his usual hyperbole):

> Each and every one of his disciples are master scientists,
> From the standpoint of publications they are authors of
> Commentaries and independent works.
> In the face of his superior attainments, Pythagoras is ashamed of his
>   shortcomings,
> And Archimedes has inevitably gone into hiding.[14]

Consequently, Taqi al-Din's plan, when presented to the Khwaja in the early months of 1575, was given immediate attention. After two or three interviews not only were Sa'adat al-Din's concerns dispelled but he became a keen supporter – his enthusiasm reflected in the fact that he arranged for Taqi al-Din to make his case in a personal audience with the sultan rather than by means of a petition.[15] Over the following year the two men became close. In 1576 Taqi al-Din observed a lunar eclipse from the roof of the Khwaja's house while also dedicating to him his treatise on sundials

The fourth figure was Sultan Murad III himself. After his audience with the chief astronomer, the sultan endorsed Taqi al-Din's request and awarded him a considerable sum: for construction of buildings, fabrication of instruments, and annual expenses (salaries of astronomers and secretaries).[16] At this point in early 1575 Murad had been on the throne for no more than a month or two, and he had not as yet revealed much of his personality or interests. However, a brief look at the sultan's twenty-one year reign reveals a fascination with astrology and esoterica, suggesting that his support of the observatory was more than the routine approval of a powerfully-backed petition. The late-sixteenth century was a time of millenarian ferment – the Grand Conjunction of Jupiter and Saturn in 1583 and the end of the first Islamic millennium in 1591 – and Murad seems to have been caught up in the excitement. Preoccupied with dream interpretation, numerology, and occult prognostication, he commissioned editions of astrological classics and scheduled the circumcision ceremony of his sixteen-year-old son, Mehmed (the most lavish production of the Ottoman period) for 1582. In the document approving Taqi al-Din's request, Murad expressed his pride at being the first Ottoman sultan to found an observatory.[17]

Work was started on the observatory – both buildings and instruments – in the early months of 1575. The entire complex – called by Taqi al-Din *The New Observatory* – was located on a ridge overlooking the Tophane quarter in the European section of Istanbul near the old French Embassy. While the large and small buildings were roofed, rectangular structures, the instruments themselves were set up in the open air, as had been in Ulugh Beg's Samarqand and would be later in Jai Singh's Shahjahanabad. Construction continued for the next two years, and Taqi al-Din and his men began observations and calculations about the end of 1577.

Our information concerning the two observatory buildings is unusually full. Although there were no ruins to be excavated (unlike Samarqand), the usual literary descriptions were supplemented by manuscript miniatures depicting the relevant personnel and activities. The Small Observatory had a tiled roof, and in the painting from 'Ala al-Din al-Mansur's history sixteen astronomers are shown at work. Niches packed with manuscripts lined the walls, and several astronomers had open volumes on their desks. Various portable instruments – quadrants, astrolabe, and clepsydras – can be seen, and the sixteen appear to be busy observing, drawing, and calculating. There was a terrestrial globe depicting Africa, Asia, Europe, and the western coast of South America, its contour reflecting the European discoveries of the late-fifteenth and early-sixteenth centuries. This building probably also housed the workshops where the larger, outdoor instruments were designed and constructed. To operate the instruments the imperial astronomers were divided into three groups: eight conducted observations, four recorded results, and four more were assistants. Despite the books and the relatively large number of astronomers, there was no mention (or picture) of students or classrooms. Unlike Ulugh Beg's institution in Samarqand, instruction did not seem to have played a part in the observatory founded by Taqi al-Din.

The Large Observatory was depicted in a miniature in *The Astronomical Instruments for the Sultan's Zīj*, an anonymous Turkish work describing the observatory's equipment. An elaborate building with brass and copper ornamentation, the large observatory contained two chimneys but no areas for observation or calculation. It seems to have functioned primarily as an administrative and residential centre.

Begun in early 1575, the observatory was substantially finished by early 1577. The two buildings were ready for occupancy and enough of the instruments had been assembled, tested, and positioned that a regular programme of observation could begin. Given the variety of instruments and the number of astronomers, a great deal must have been accomplished in the three years (1577–80) of the institution's existence. Taqi al-Din's goal, and the *raison d'etre* of the institution, was to update and correct Ulugh Beg's *Zij al-Sultani*, and the evidence from the chief astronomer's two *zijs* reveals that he had begun his task (with portable instruments) as early as 1573, two years before the foundation of the observatory. Examples of the observations made during the institution's three-year life include: in 1577 two solstices and an equinox, and in 1579 an equinox and two solar sightings. But the most important observation of the institution's short life was the bright comet of November 1577. Appearing in the Istanbul sky on the first night of Ramadan, it flared over the capital for the next forty days. Recognising this as a rare opportunity, Taqi al-Din readied an interpretation. Since the Ottomans had just launched a campaign against the Safavid rulers of Iran, the chief astronomer stressed, in his audience with Murad III, the military implications of the comet. Both the tail and the head of the comet pointed east toward Iran, as if to discharge their ominous fire on the Safavids. In addition, the comet first appeared in the house of Sagittarius, symbolising the Ottoman archer, and it would disappear in Aquarius, a sign of peace and prosperity awaiting the archer. In conclusion, al-Mansur has the astronomer saying:

> Oh, world swaying king!
> The candle of your pleasant society shall be resplendent.
> There are joyful tidings for you concerning the conquest of Persia,
> For the foe is lying, with failing breath, upon the ground.
> [This] Is for this realm an indication of well being and splendor,
> But for Persia it is a bolt of misfortune,
> And its guide is the Tradition that 'wickedness is from over there' [i.e., from the East],
> As he had made a pleasing forecast for the occasion,
> He received kindness and benefits from the King of the World.[18]

And the Ottoman victory over the Safavids at Tiflis (capital of the Caucasus) early in the following year seemed to validate the accuracy of Taqi's interpretation.

The instruments of the Istanbul Observatory have been thoroughly described and depicted – in the prose of Taqi al-Din's Arabic *zij* (*Culmination of Thoughts in the Kingdom of Rotating Spheres*), in the poetry of 'Ala al-Din al-Mansur's Persian history (*Shahinshahnama*), and in the prose and miniatures of the anonymous author's Turkish inventory (*The Astronomical Instruments for the Sultan's Zij*).[19] They can be divided into three groups: those inherited from the early Greek or Indian astronomers, those invented by earlier Islamic astronomers, and the entirely new creations of Taqi al-Din and his men.[20]

In the first group are the armillary sphere, the triquetrum or parallactic ruler, and the diopra. Taqi al-Din's armillary sphere was unusually large – six rings with radii of four metres each and its measurements were uncommonly accurate.[21]

> People of discernment found out with the help of the armillary sphere
> All the positions of stellar bodies in latitude and longitude.[22]

The triquetrum or parallactic ruler was also an ancient instrument. The one in the Istanbul observatory was made from three long pieces of wood – a perpendicular graduated scale and two pivoted arms hinged at the top and bottom. It was used to determine the altitude and zenith of the Moon.[23]

> They also took measurements
> With the parallactic rule . . . all angles of elevation . . .
> And also the parallax of the moon were determined.[24]

The dioptra was a sighting instrument. At the Istanbul Observatory it was a rod with circles or holes at each end and was used to measure the apparent diameters of stars, planets, and eclipses.[25]

> They also took measurements . . .
> With the . . . diopter; . . . by the help of this instrument . . .
> The dimensions and distances of the stellar bodies were recorded in an orderly
>   manner.[26]

The second group of instruments, those invented by Islamic astronomers, included the mural quadrant, the azimuthal semicircle, and the corded instrument. Taqi al-Din's mural quadrant was an angle-measuring device mounted on a wall and precisely aligned to the observatory's meridian. Comprised of two brass quadrants (radii of six metres each), it measured the altitude of the Sun and stars. The Ottoman astronomer preferred his instrument to the mural sextant of Ulugh Beg, probably because the quadrant could measure a full 90-degree arc whereas the sextant covered only 60 degrees.[27]

> With the mural quadrant the declination of the sun was ascertained
> And other distances from the equator were also determined.'[28]

The azimuthal semicircle was used to determine the altitude and azimuth of stars. Taqi al-Din's instrument consisted of a copper ring representing the horizon and a semicircle perpendicular to the ring.[29]

> With the help of the instrument for measuring azimuths and altitudes,
> Angles of elevation were recorded by astronomers who worked together.[30]

The instrument with chords was used to determine the fall and spring equinoxes. It consisted of a rectangular base with two cords attached.[31]

> The corded instrument was by no means in the background either,
> For with its help the points of the equinoxes were correctly determined.[32]

The third group of astronomical instruments were those invented by the Ottoman astronomers and included the sextant, the wooden quadrant, and the astronomical clock. Taqi al-Din's sextant consisted of three rulers – two attached like the hinged arms of the triquetrum and the third an arc. It was designed to measure the angle between two stars, enabling the astronomer to determine the distance between them.[33]

> With the help of the [sextant], And thanks to very carefully made observations,
> The radius of Venus' epicycle, in the third firmament,
> Became known with great precision.[34]

The wooden quadrant, like the larger mural quadrant, was used to measure the altitude of stars. Consisting of three rulers, it formed a quarter circle, and could measure angles up to 90 degrees. Its advantage over the larger instrument was its portability. The astronomer could make observations at different times and places and in different weather.[35]

> The intricate and complex aspects of the motions of Mercury and Venus,
> Which are caused by the revolution of the aged ecliptic,
> As well as the angles of elevation and the zenith distances
> Were checked and confirmed with the ruler-quadrant.[36]

The observational clock, described in Taqi al-Din's treatise of 1559, was used to determine the right ascension of stars. This clock, the most accurate of its day, measured time in hours, minutes, and seconds.[37]

> And with the help of careful measurements and corrections with the clock,
> The ascensions of the stellar bodies were fixed.[38]

With it the Ottomans were able to more accurately and precisely determine the location of stars – the mural quadrant measuring their declination, degrees above the equator, the equivalent of terrestrial latitude; and the observational clock measuring their right ascension, the distance in hours, minutes, and seconds east of the vernal equinox, the equivalent of terrestrial longitude.[39]

With his new observatory – equipped with advanced instruments and staffed by experienced astronomers – Taqi al-Din was able to make a good deal of progress on his goal of updating Ulugh Beg's astronomical tables. Although he and his men were limited to no more than seven years of observations (1573–80), they were able to substantially complete a new zij – *Culmination of Thoughts in the Kingdom of Rotating Spheres* also known as *The Emperor's Zij*. Taqi al-Din's composition lacked a traditional conclusion but it contained the most accurate set of astronomical tables in the Eurasian world. This was the result of two factors: the precision of the new instruments and the superiority of the new mathematics. Consisting of some forty-seven folios, the *zij* opened with forty pages of trigonometric calculations, followed by material on astronomical clocks, celestial bodies, and observational instruments. Taqi al-Din's work was groundbreaking. He was responsible for the most accurate contemporary calculation of the obliquity of the ecliptic – his value of 23 degrees 28 minutes and 40 seconds was quite close to the modern determination of 23 degrees 26 minutes and 21 seconds. He also obtained more precise estimates for several solar parameters. Using the new method of three points – triangulating on Venus, Aldaberan (the brightest star in the constellation of Taurus), and Spica Virginis (located near the ecliptic) – he calculated the eccentricity of the Sun as two degrees and the annual movement of the Sun's apogee as 63 seconds. Quite close to the modern figure of 61 seconds, Taqi al-Din's result was much more accurate than those of either Copernicus (24 seconds) or Tyco Brahe (45 seconds). In the measurement of angles he used trigonometric functions – sine, cosine, tangent, cotangent – rather than the traditional chord technique. While Copernicus and the early Europeans still employed Ptolemy's interpolation method, Taqi al-Din's use of trigonometry yielded much more accurate results. In this *zij*, however, he expressed his results in the sexagesimal number system rather than in the decimal.[40]

As was the case with the other famous Islamic *zijs* – those of Ulugh Beg and Nasir al-Din Tusi – other versions of Taqi al-Din's *Culmination* soon appeared. The first – *Non-Perforated Pearls and Roll of Reflections* – carried a date of 1581–2, emerging no more than two or three years after the completion of the *Culmination*. Dedicated to Khwaja Sa'adat al-Din, Murad's former tutor and an early supporter of the observatory, this treatise was a revision of the original, substituting the latitude of Cairo for that of Istanbul. Its significance, however, lay in the fact that it was the first *zij* to express trigonometric functions in decimal fractions rather than in the traditional sexagesimal number system.[41]

The third early version of Taqi al-Din's *zij* was *A Simplification of the Emperor's Zij*. A shorter and condensed version of the original, this treatise employed the decimal system for all astronomical and astrological information. Only the star tables remained in the traditional sexagesimal format. One of Taqi al-Din's most important innovations was popularising the use of decimals.[42]

On 22 January 1580 Murad III ordered Admiral Kilic 'Ali Pasha to raze the Observatory. 'Ala al-Din Mansur wrote:

> Upon the orders of the exalted Sovereign.
> They pulled the Observatory up by the roots
> And made to subside all the work concerning the firmament.
> The armillary sphere was uprooted from its foundation,
> And the instruments were broken and the nails pulled out. Nothing remained of
>   the Observatory but name and memory;[43]

The reason given for the destruction was religious – the presumption of the mathematicians and astronomers. 'Ala al-Din Mansur explained:

> Do not make decisions concerning the affairs of the firmament.
> For who, besides God, knows the gait and the revolution of the heavens? . . .
> When the affair concerning the Observatory was brought to completion,
> And it was torn from its foundation and its traces were obliterated,
> All people of faith prayed for the mighty King;
> For he had caused the performance of a deed which was in accordance with the
>   Law of the True Religion.[44]

Although the historian stressed the religious aspect, other mostly political factors probably played a role as well. In fact, Murad III, far from punishing Taqi al-Din, consulted the chief imperial astronomer before making his final decision.

At this time, all of a sudden, the Potentate who is the Defender of the Religion Spoke thus to his astronomer, Taqi al-Din:

> People of learning have made inquiries concerning this:
> Oh you witty man of conscientiousness and perfection
> Inform me once more on the progress and the results of observations.
> Have you entangled knots from the firmament in a hairsplitting manner?

Taqi al-Din answered:

> In the Zij of Ulugh Bey
> There were many doubtful points, oh exalted King;
> Now through observations the tables have been corrected, . . .
> From now on, order the abolishment of the Observatory,
> To the consternation of the ill-wishers and the jealous.[45]

In fact, Taqi al-Din and his observatory seem to have been the victims of a power struggle at the highest levels of the Ottoman state. Sokullu Mehmed Pasha had been appointed Grand Vizier in 1565, a year before the death of Suleiman the Magnificent, and had become increasingly dominant during the reign of the weak Selim II (1566–74). And barely a month after the accession of Murad III, Taqi al-Din, with the support of Sokullu and Khwaja Sa'adat al-Din, had convinced the new sultan to underwrite the construction of an observatory. Soon after his accession Murad III became dissatisfied with his subordinate position and began a campaign (with the assistance of his mother, wife, and the Shaikh al-Islam) to weaken Sokullu and regain the reins of government. In 1577 and 1578 he reassigned or removed several of Sokullu's most important allies. And in October 1579 Sokullu himself was assassinated – perhaps by an agent of the sultan's wife. Three months later the order to abolish the observatory was given. The Shaikh al-Islam had provided religious reasons for his opposition, but given Murad's interest in astrology and his pride in the observatory, its destruction seems more a by-product of a political skirmish than a triumph of religious fanaticism.[46]

Taqi al-Din was 54 when the observatory was destroyed, and he died five years later in 1585. Despite 'Ala al-Din Mansur's statement concerning the chief astronomer's acquiescence, it is not difficult to imagine the deep disappointment he must have felt at the tragic end to his state-of-the art institution, the opportunities for research and publication forever lost. Broken-hearted or not, Taqi al-Din continued to work during the five years he had left. While it is impossible to date all of his publications, many of the mathematical and astronomical works probably appeared during this period. He wrote an introduction to Indian mathematics and a groundbreaking treatise on extracting the values for chord 2 degree and sine 1 degree. These problems had occupied al-Kashi and Ulugh Beg earlier and Taqi al-Din came up with a new, more accurate solution which he expressed in the decimal rather than the sexagesimal number system.[47] In astronomy he composed a number of treatises (33 all told). In addition to the three *zij*s, he wrote on the conversion of dates from one calendar to another, finding the *qibla* direction, the differences between the real and false horizon, and the astrolabe as a timekeeping instrument.[48] Like the other polymaths, his interests were wide-ranging. He compiled an alphabetical inventory of medicines and a work on zoology. He also composed a treatise on optics. *Book of the Light of the Truth of the Sights* was dedicated to Murad III and was the most comprehensive treatment of the topic in the Islamic world since the *Book on Optics* of Ibn al-Haytham (965–1039). It was divided into three parts. Part One on direct vision: the structure of the eye, the nature of vision, and the effect of light on sight. Part Two on reflection: reflected light, variations due to mirrors, and errors of sight due to reflection. Part Three on refraction: variations of light while travelling through mediums of different densities.[49]

Notes

1. Stephen P. Blake, *Time in Early Modern Islam: Calendar, Ceremony, and Chronology in the Safavid, Mughal, and Ottoman Empires* (New York: Cambridge University Press, 2013), 37–43.
2. Ihsanoglu, *History of the Ottoman State*, 2: 375–6.
3. For a discussion see Blake, *Time*, 69–70.
4. Ihsanoglu, *History of the Ottoman State*, 2: 407–8.
5. Ibid. 2: 409–10.
6. Ibid. 2: 530–32; Ilay Ileri, "Ali al-Qushji and His Contributions to Mathematics and Astronomy', *Journal of the Center for Ottoman Studies* 20 (2006): 175–83.
7. Salim Ayduz, 'Bio-bibliographical Essay', <http://www.muslimheritage.com> last accessed 21 July 2008; Salim al-Hassani & Mohammed A. al-Lawati, 'The Six-Cylinder Water Pump of Taqi al Din: Its Mathematics, Operation, and Virtual Design', <http://www.muslimheritage.com> last accessed 21 July 2008.
8. For a discussion see Blake, *Time*, 66–9.
9. Salim T. S. al-Hassani, 'The Astronomical Clock of Taqi al Din', <http://www.muslim

heritage.com> last accessed 21 July 2008; Sevim Tekeli, *The Clocks in Ottoman Empire in 16th Century and Taqi al Din's 'The Brightest Stars for the Construction of the Mechanical Clocks'* (Ankara: T. C. Kültür Bakanliği, 2002).

10. Aydin Sayili, "Ala al-Din al-Mansur's Poems on the Istanbul Observatory', *Belleten* 20 (1956): 473.
11. Ibid. 471.
12. Ibid. 472.
13. Ibid. 474.
14. Ibid. 475.
15. Ibid. 476.
16. Ibid.; Sayili, *Observatory*, 302–3.
17. Blake, *Time*, 63–73, 98.
18. Sayili, 'al-Mansur's Poems', 290–1.
19. Sevim Tekeli, 'The Observational Instruments of the Istanbul Observatory', <http://www.muslimheritage.com> last accessed 21 July 2008.
20. Ibid.
21. Ibid. 3.
22. Sayili, 'al-Mansur's Poems', 478.
23. Tekeli, 'Observational Instruments', 6.
24. Sayili, 'al-Mansur's Poems', 478.
25. Tekeli, 'Observational Instruments', 7.
26. Sayili, 'al-Mansur's Poems', 478.
27. Tekeli, 'Observational Instruments', 4.
28. Sayili, 'al-Mansur's Poems', 478.
29. Tekeli, 'Observational Instruments', 5.
30. Sayili, 'al-Mansur's Poems', 478.
31. Tekeli, 'Observational Instruments', 4.
32. Sayili, 'al-Mansur's Poems', 478.
33. Tekeli, 'Observational Instruments', 7–8.
34. Sayili, 'al-Mansur's Poems', 478.
35. Tekeli, 'Observational Instruments', 7.
36. Sayili, 'al-Mansur's Poems', 478.
37. Tekeli, 'Observational Instruments', 9.
38. Sayili, 'al-Mansur's Poems', 479.
39. Tekeli, 'Observational Instruments', 12.
40. Sayili, 'al-Mansur's Poems', 438–9; Ihsanoglu, *History of the Ottoman State*, 2: 412, 554–5; Ihsan Fazioglu, 'Taqi al Din Ma'aruf: Survey on Works and Scientific Method', <http://www.muslimheritage.com> last accessed 21 July 2008; Sevim Tekeli, 'Taqi al Din Ma'aruf's Work on Extracting the Cord 2 Degree and Sine 1 Degree', <http://www.muslimheritage.com> last accessed 21 July 2008.
41. Ihsanoglu, *History of the Ottoman State*, 2: 412, 554–5; Fazioglu, 'Taqi al Din Ma'aruf: Survey on Works and Scientific Method'.
42. Fazioglu, 'Taqi al Din Ma'aruf: Survey on Works and Scientific Method'.
43. Sayili, 'al-Mansur's Poems', 482–3.
44. Sayili, 'al-Mansur's Poems', 483–4.
45. Ibid. 481–2.
46. Ihsanoglu, *History of the Ottoman State*, 2: 413; Sayili, *Observatory*, 302–3.
47. Sevim Tekeli, 'Taqi al Din ibn Ma'aruf's Work on Extracting Cord 2 degree and Sine 1 Degree'.

48. Salim Aydoz, 'Manuscripts of Taqi al Din's Works', <http://www.muslimheritage.com> last accessed 21 July 2008.
49. Huseyin Gazi Topdemir, 'Taqi al Din Ma'ruf and the Science of Optics: The Nature of Light and the Mechanism of Vision', <http://www.muslimheritage.com> last accessed 21 July 2008.

CHAPTER EIGHT

The observatory in Shahjahanabad

Islam reached the Indian subcontinent soon after the death of the prophet. The earliest Muslims were Arab traders, who landed on the south-western coast of India in the late seventh century CE. In 711 the Arab general Muhammad Qasim invaded from the northwest and, defeating the local raja, established Sind as the easternmost province of the Umayyad Caliphate. For the next three centuries Islam remained a curiosity on India's northern fringes. In the chaos attending the collapse of the Abbasid Caliphate in the late tenth century, a series of Central Asian military tribal confederations began to rule Afghanistan. Drawn from the Turkish nomads of the northern steppes, these horseback warriors formed the ruling aristocracy of the Islamic states that would dominate north India for the following seven hundred years.

The Ghaznavids (977–1186), under the leadership of Mahmud of Ghazni, conquered much of the Punjab, establishing their capital in Lahore. They were followed by another dynasty of Turkish warriors, the Ghurids (1186–1215), who under Qutb al-Din Aibak (1206–10), occupied Delhi. Qutb al-Din founded the first in a series of dynasties known collectively as the Delhi Sultanate (1206–1526). What these dynasties had in common was the ethnicity of their rulers – Afghan or Turkish military elites from Central Asia. Under 'Ala al-Din Khalji (1296–1316) these Muslim warriors extended their control over Gujarat, Rajasthan, the Deccan, and parts of south India. The Khaljis were followed by the Tughluqs (1320–1413). Muhammad ibn Tughluq (1325–51), founder of the mature state, strengthened his rule by recruiting converts and the newly-immigrated, counterbalancing the power of the long-established aristocratic families. The ruling elite of the Delhi Sultanate were a religious and cultural minority, aggressively Muslim in the largely non-Muslim environment of the subcontinent. These warriors pledged allegiance to the Caliph, supported the judicial authority of the *ulama*, and welcomed talented newcomers from the towns and cities of eastern Islam.[1]

Abu Rayhan Muhammad ibn al-Biruni was one of the first Islamic astronomers to explore Indic astronomy/astrology. Although the astronomers of al-Ma'mun's House of Wisdom had translated the works of Aryabhatta and Brahmagupta into Arabic in the early ninth century, they had not visited the subcontinent. In his conversations with the Sanskrit scholars, al-Biruni absorbed a good deal of the indigenous theories while also passing on the latest work of the Islamic astronomers. With the establishment of the Delhi Sultanate, scholars and poets joined soldiers and administrators at the Indian courts. The first Indian *zij* was compiled by Mahmud ibn 'Umar at the court of Sultan Nasir al-Din Shams al-Din Iltutmish (1246–65). Although the *Zij-i Nasiri* was roughly contemporary with the *Zij-i Ilkhani* of Nasir al-Din Tusi, there is no evidence that Mahmud ibn 'Umar had access to an observatory or made fresh observations. In all likelihood, the *Zij-i Nasiri* (no copy remains) was an updated version of an earlier treatise – perhaps the *Zij-i Malik Shahi* of 'Umar Khayyam.[2]

Under the Khaljis, and especially during the reign of 'Ala al-Din Khalji, the demand for skilled astronomers swelled, and a large number of specialists arrived in the subcontinent, casting horoscopes, compiling almanacs, and bringing news of the latest mathematical and astronomical discoveries. The response to this stimulus was widespread. The Khalji poet Amir Khusraw (1253–1325), for example, composed a poem on the twenty-eight lunar mansions called *Manazil (Houses)*.[3] And work on the first Indian observatory was begun in 1407 in Balaghat by order of Sultan Firuz Shah Bahmani (1397–1422). Unfortunately, its designer, Hakim Hasan Jalani, died soon after ground breaking, and the observatory was never completed.[4]

The first extant Indian *zij* was the *Zij-i Jami Mahmud Shah Khilji*. Dedicated to Mahmud Shah Khilji (1435–69), a ruler of the Malwa Sultanate (1436–1531), it was the effort of an anonymous astronomer who was commissioned by a high-ranking noble to prepare an almanac. The almanac was lost in a fire, and the astronomer was asked to compile an astronomical treatise instead, which he dedicated to the reigning sultan. The *zij* consisted of an introduction, two chapters, and a conclusion but the only surviving copy is defective – the first pages and the entire second chapter are missing. The introduction contained thirty six sections: the first consisted of definitions (astronomy and geometry), sections two to sixteen covered arithmetic, section seventeen was on measurement, sections eighteen to twenty-four treated sexagesimal arithmetic, and the last twelve sections concerned the astrolabe. The first chapter had twenty-two sections: the first on terminology, sections two to five on the Hijra, Roman, Persian (Yazdegird), and Maliki calendars and eras, section six on the conversion of dates from one calendar to another, sections seven to nine on the Turkish Twelve-Year Animal Calendar, section ten on the motions of the Sun, section eleven on the motions of the Moon, and sections twelve and thirteen on the motions of the five wandering planets. Section fourteen took up planetary conjunctions and oppositions, sections fifteen to seventeen were devoted to planetary motions, sections eighteen and nineteen examined astrological problems, and sections twenty and twenty-one dealt with lunar and solar eclipses. In the last section the author warned that astrological predictions were unreliable; the only predictions that could be trusted were those based on arithmetical computations. He also declared his admiration for the *Zij-i Ilkhani* of Nasir al-Din Tusi.

In the *Zij-i Jami Mahmud Shah Khilji* the author covered much the same ground as had Tusi – the only additional materials coming from Tusi's *Thirty Chapters* and his *Treatise on the Science of Astronomy*.[5]

Indian astronomers were also interested in instrumentation. Al-Biruni wrote extensively on the astrolabe, and the later immigrant astronomers deemed it an indispensable tool of their trade. In the late fourteenth century Firuz Shah Tughluq (1351–88), successor to Muhammad ibn Tughluq, funded an astrolabe workshop. He also commissioned manuals (in both Persian and Sanskrit) on astrolabe construction and operation. The Sanskrit treatise (1370) by the Jain monk Mahendra Suri was the only one to survive.[6]

The last Muslim dynasty of the Indian subcontinent was the Mughal (1526–1739). Like the rulers of the Delhi Sultanate, the Mughal emperors were also descendants of Turkish military aristocrats. Babur, the founder of the dynasty, was the great-great-great grandson of Timur. Like the Ottomans, the Mughals played a vital role in the development of Islamic science. Humayun, Babur's son, demonstrated a wide-ranging interest in astronomy and astrology. Not only did he support a sizeable stable of household astronomers (both Muslim and Indic), but he developed a considerable expertise himself. Although not as talented as Ulugh Beg, he was not a dilettante. The historian Abu Fadl wrote of Humayun:

> His Majesty . . . in astrolabic investigations and studies in astronomical tables and observations was at the head of the enthroned ones of acute knowledge . . . He had . . . extraordinary excellence in the astrolabe, globe and other instruments of the observatory.[7]

Mulla Chand, one of Humayun's astronomers, cast horoscopes for the imperial household. He had accompanied Humayun on his flight from India to Iran in 1530, where the Mughal emperor had sought refuge with the Safavid ruler Shah Tahmasp (1524–76). Since Hamida Begum, Humayun's wife, was about to give birth, the emperor left her and Mulla Chand in the fort of Amarkot. As a result, the astronomer was able prepare a horoscope (with the exact time of birth) for the new-born child – Humayun's son and successor Akbar.

Humayun also sponsored a productive and long-lasting family of astrolabe makers from Lahore who, over four generations, signed their names to 100 astrolabes and twenty-five celestial globes. The patriarch of the family was Ustad Shaikh Allahdad Asturlabi Humayuni Lahuri. He constructed the oldest extant astrolabe in India (1567), and his son Mulla 'Isa produced five astrolabes between 1600 and 1604. Mulla 'Isa had two sons: Qa'im Muhammad was responsible for four celestial globes and five astrolabes, and Muqim constructed thirty-eight astrolabes between 1609 and 1659. Qa'im's son, Diya al-Din Muhammad, was the most prolific of all, producing thirty astrolabes and sixteen celestial globes between 1645 and 1680.[8]

The final manifestation of Humayun's passion for the heavenly sciences was his decision (like Ulugh Beg's) to build an observatory. He chose a site and collected instruments, but in a cruel irony he fell from an observation tower while awaiting the appearance of the planet Venus and died. The project was cancelled.[9] Humayun's demise, however, did not spell the end for astronomy and astrology in Mughal India. On the contrary, it was during the reign of his son Akbar that the study of the rational sciences became compulsory in state madrasas. This decision, promoted by Mir Fathullah Shirazi (d. 1589), the Safavid polymath who arrived at Akbar's court in 1583, made astronomy, mathematics, medicine, and philosophy part of the everyday curriculum. In addition to the state institutions, a great deal of scientific education also took place in the private residences of eminent scholars. For example, five students of the famous Iranian philosopher Jalal al-Din Dawani (1426–1502) came to the subcontinent and started a school for the rational sciences. Dawani, an important member of the Mansuriyya madrasa in Shiraz, was best-known for his *Akhlaq-i Jalali*, an ethical treatise in the style of Tusi's *Akhlaq-i Nasiri*. In India, however, his students specialised in mathematics, astronomy, and logic. One of them, Abu al-Fadl Gazruni, was famous for drilling his students on Ptolemy's *Almagest* and Tusi's *Treatise on the Science of Astronomy*.[10]

Mir Fathullah Shirazi was another polymath. He taught Shahjahan's court astronomer, Mulla Farid Ibrahim Dihlavi (d. c. 1631), and had Ulugh Beg's *Zij al-Sultani* translated into Sanskrit. And he was almost surely the source for the eighty-six *zij*s listed in Book Three of Abu Fadl's *Ain-i Akbari*. The only new *zij* of Abkar's reign, however, was completed by Mulla Chand in the years before Shirazi's arrival. *Tahsil Zij-i Ulugh Beg* was, as its title indicated, a simplified version of Ulugh Beg's *Zij al-Sultani*. Mulla Chand retained the basic three-chapter arrangement of the original – leaving out the chapter on astrology and the star tables. Chapter One was on chronology, and Mulla Chand dropped the material on the Twelve-Year Animal Era and Calendar but left the other eras and calendars unchanged. Chapter Two covered trigonometry and spherical astronomy. To this chapter the astronomer added two new sections while rearranging several others. Chapter Three of Ulugh Beg's treatise was on planetary motion and Mulla Chand left it untouched.

In the astronomy and astrology of the subcontinent, however, Mir Fathullah's major accomplishment was the design and implementation of a new solar era. Akbar's *Tarikh-i Ilahi* (*Divine Era*) was introduced 21 March 1584, and its epoch, following Iranian custom, was the vernal equinox following Akbar's accession – 21 March 1556. Its calendar was solar also, employing the month and day names of the Jalali calendar.[11] The first new era in the Islamic world since the *Tarikh-i Jalali* of 'Umar Khayyam in the eleventh century and the Turkish Twelve-Year Animal Era in the twelfth and thirteenth centuries, Akbar's *Tarikh-i Ilahi* was a significant but controversial addition to the temporal system of Mughal India.[12]

Although 'Umar Khayyam's era had been, for the most part, an attempt to divert Malik Shah's attention (the ruler had wanted a complete revision of the *Almagest* [a thirty-year undertaking]), it had also appealed to the Seljuq ruler's ego: his name would live forever. For Mir Fathullah and Akbar, on the other hand, the motivation was more prosaic. The new solar era was shorter and simpler – the current eras had all passed 1,000 – and would make the work of administrators and merchants easier. In addition, the *Ilahi*, like the *Jalali*, would make the collection of agrarian taxes fairer. Dishonest officials would no longer be able to manipulate lunar and solar calendars in order to cheat the peasants. However, Akbar's new solar era, like Taqi al-Din's new observatory, ran into opposition. During the reign of Akbar's grandson Shahjahan (1628–58) the orthodox *ulama*, threatened by the influx of Indic customs and traditions, demanded a revision of the first volume of his official history, declaring its chronological framework – the *Tarikh-i Ilahi* era and calendar – heretical. As a result, in 1638 Shahjahan was forced to reject Mirza Amin Qazwini's chronicle of his first decade and to appoint another historian, 'Abd al-Hamid Lahori, to draft a new version using the traditional Hijra era.[13]

Although Islamic astronomer/astrologers became increasingly important in the lives of Indian Muslims (both immigrants and converts), the Indic practitioners (*joytish*) were never entirely displaced. Continuing to serve the non-Muslim majority population, they were often included in the households of wealthy and powerful Muslims as well. Mughal emperors, for example, had two horoscopes (Indic and Islamic) drawn up at the birth of their sons, and both specialists were consulted when mulling over important decisions – constructing a palace, contracting a marriage, planning a campaign, or organising a journey.[14]

Akbar, for example, patronised Indic astronomers – the Hindus Muni Sundar and Nilakantha Anata Chintamani and the Jain Padmasundara. Nilakantha was the most important of the three, serving as Akbar's *Royal Astrologer* (*Jyotish-Raja*). He composed the sections on astronomy and astrology in the vast Sanskrit encyclopaedia (*Todarananda*) and finished his own astrological treatise (*Tajikanilakanthi*) in 1587. In 1590 Nilakantha's brother, Rama, compiled an astronomical handbook (*Ramavinoda*) for an official at the Mughal court; its epoch was 1556, the first year of Akbar's reign. Jahangir, Akbar's son and successor, consulted Shrikrishna, Vishwanath Diwakar, and Keshava Sharma, who was given the title of 'chief astronomer' (*jyotish-rai*). Under Shahjahan, the Indic astronomer Nityananda translated Mulla Farid's *Zij-i Shahjahani* into Sanskrit as the *Siddhantasindhu*. Even the ultra-conservative Aurangzeb included the Indic practitioners, Ishvar Das and Manirama Kidshita, among his household astronomers.[15]

Shahjahan, Akbar's grandson, continued the Mughal commitment to astronomy and astrology. Like his great grandfather, he contemplated the construction of an observatory. He commissioned a plan by Mulla Mahmud Jawnpuri but, lacking sufficient funds, was never able to break ground.[16] His court astrono-mer, Mulla Farid had in 1597 joined the household of 'Abd al-Rahim Khan-i Khanan (1556–1626), one of Akbar's high-ranking nobles. Khan-i Khanan was passionate about astrology, having written two books on the subject, and Mulla Farid dedicated an astronomical treatise, *Siraj al-Istikhraj*, to his patron.[17] Much later, after Khan-i Khanan's death, Mulla Farid came to the attention of Asaf Khan (d. 1641), chief minister of Shahjahan and father of Mumtaz Mahal (Shahjahan's queen and the inspiration for the Taj Mahal). Asaf Khan wanted to immortalise the reign of his son-in-law by inaugurating a new era – like the *Tarikh-i Jalali* of Malik Shah or the *Tarikh-i Ilhai* of Akbar. Shahjahan approved of the idea, and Mulla Farid was asked to produce a new *zij* – using the new era. The *Zij-i Shahjahani* was presented to the emperor two years later, and Mulla Farid died soon after.

Because there were neither instruments nor time for new observations, the *Zij-i Shahjahani* (1629–30) was mostly a revision of the *Zij al-Sultani*. In his intro-duction, Mulla Farid distinguished two kinds of astronomical treatises:

Zij-i Rasadi or observational treatise [and]
Zij-i Hisabi or computational treatise.[18]

Since constructing, equipping, and staffing an observatory was a lengthy and expensive undertaking, most *zijs* were computational. That is, they involved revising and updating the parameters of a previously completed observational treatise. The value of this *zij* depended on the extent of the revision. In early-modern India, the most recent observational treatise was that of Ulugh Beg, and, as a result, Mulla Farid chose it as his starting point.

The *Zij-i Shahjahani* was divided, like the treatises of Tusi and Ulugh Beg, into an introduction and four chapters. The introduction contained five sections: the first on definitions, the second on special features, the third on numbers, the fourth on corrections and additions to the *Zij al-Sultani*, and the fifth on the month, day, and year. The first chapter was on eras and calendars and had ten sections: the first on the Ilahi Era, the second on the Hijra, the third on the Greek, the fourth on the Yazdegird, the fifth on the Jalali, and the sixth on conversion. The seventh section was on the Indic Samvat Era, the eighth on conversion from the Hijra to the Samvat calendar and era, the ninth on the Turkish Twelve-Year Animal Era, and the tenth on important festivals.

The second chapter took up the determination of planetary ascendants (trigonometry and spherical trigonometry), and the third chapter covered planetary positions and motions. While the text for these two chapters was taken almost verbatim from the *Zij al-Sultani*, many of the tables were updated. The fourth chapter dealt with astrology. Because of the Indian material and the revisions and updates, the *Zij-i Shahjahani* was an obvious improvement. Shahjahan ordered a Hindi translation for the general public, and perhaps the court historian did not overly exaggerate when he observed that the new treatise relegated Ulugh Beg's *zij* to the dustbin.[19]

In the Islamic world, the last observatory (or rather observatories) and the last astronomical treatise were created by a Rajput ruler for the Mughal emperor Muhammad Shah (1719–39). Maharajah Jai Singh (1688–1743), Hindu ruler of the Raput state of Amber and high-ranking Mughal officeholder, constructed five observatories in North India between 1721 and 1738. He also compiled the *Zij-i Jadid-i Muhammad Shah* (*The New Astronomical Treatise of Muhammad Shah*) in 1731–32. The result was curiously old fashioned: the observatories were obsolete (lacking proper telescopes), and the treatise was out of date (employing the geocentric Ptolemaic cosmology rather than the heliocentric cosmology of Copernicus and Kepler). Nevertheless, Jai Singh's creations were instructive because they illustrated the altered relationship between the Islamic and the European scientific worlds. By the mid-eighteenth century leadership in astronomical theory and instrumentation had passed from the Islamic East to the European West.

Like Ulugh Beg, Jai Singh displayed an early passion for mathematics and astronomy. According to the *Zij-i Jadid Muhammad Shahi* '. . . from the first dawning of reason . . . [he] was entirely devoted to the study of mathematical science (astronomy).'[20] At age thirteen Jai Singh acquired two treatises on astronomy, and soon after he met and befriended Jagannatha Samrat, the Indic astronomer who was to become his life-long teacher, advisor, and collaborator. In the turbulent years following the death of the Aurangzeb (1658–1707), Jai Singh developed into a rich and powerful figure at the Mughal court. He controlled a large part of central Rajasthan and Gujarat and had the resources to not only support an army and administration but to also indulge his passion for astronomy and astrology.[21]

In November 1720, soon after his accession, Muhammad Shah invited Jai Singh to Shahjahanabad, his capital. In addition to pledging his loyalty to the new ruler, Jai Singh brought up the matter of a new observatory. According to the *Zij-i Jadid-i Muhammad Shahi*:

> He [Jai Singh] found that the calculation of the stars as obtained from the tables in common use, such as the new tables of Sayyid Gurgani [*Zij al-Sultani* of Ulugh Beg] and Khaqani [*Zij-i Khaqani* of al-Kashi] and the Tashilat-i Mulla Chand Akbar Shahi [Mulla Chand's *Tahsi-i Ulugh Beg*] and the Hindu books and the European tables . . . give . . . widely different results than those determined by observation . . . especially with the appearance of the new moon . . . [also] the time of the rising and setting of the planets, and the seasons of the eclipses of the sun and the moon . . .
>
> He [Muhammad Shah] was pleased to reply, since you, who are learned in the mysteries of science and have a perfect knowledge of this matter, having assembled the astronomers and geometricians of the faith of Islam, the Brahmins and

Pundits, astronomers from Europe, and having prepared all of the apparatus of an observatory . . . you [should] labor for the ascertaining . . . that the disagreement between the calculated times of these phenomena and the times in which they are observed to happen may be rectified . . . to accomplish the exalted command . . . he [Jai Singh] constructed here [at Shahjahanabad] several of the instruments of an observatory, according to the books of the Islamic school of astronomy such as the ones erected at Samarqand.[22]

In Shahjahanbad, as in Samarqand and Maragha, founding an observatory meant fabricating instruments (deployed out-of-doors) in addition to erecting buildings. Initially, the Raja had wanted to construct the traditional Islamic instruments out of metal. He assembled six of brass (the dioptra, triquetrum, armillary sphere, revolving parallactic ruler, and 60-degree meridian sextant) but found that they were inaccurate. Either they were so small that their scales could not be divided into minutes or they were so big and heavy that their axes began to wear down, displacing the centre and shifting the planes of reference. As a result, Jai Singh decided to build large masonry instruments of his own design: his goal was to achieve an accuracy of one minute of arc in the measurement of angles and two seconds in the measurement of time.[23]

In early 1721 Jai Singh's men began erecting six instruments in an open area outside the walls of the Mughal capital. This group of instruments came to be known as Jantar Mantar or Calculation Instrument; the same name was also given to the collection of instruments in both Jaipur and Ujjain. As in the earlier Islamic observatories, the six Shahjanabad instruments were either time-measuring or angle-measuring devices. The most important and imposing instrument in the Shahjahanabad Jantar Mantar was the Samrat Yantra or Supreme Instrument. A large equinoctial sundial, its gnomon was 21.3 metres tall and its hypotenuse 39 metres long. The quadrants had a radii of 15.09 metres and were 2.65 metres wide. A flight of stairs led to the top. The great size of the instrument was intended to provide greater accuracy. The quadrants were divided into hours, minutes, and seconds. The hours were subdivided into 30, 15, 5, 1, and ½ minute markings; each half-minute was further subdivided into fifteen markings of two seconds each – indicating the level of accuracy expected.[24]

The Sasthamsa Yantra was a sixty degree meridian sextant used to measure the declination, zenith distance, and diameter of the Sun. Like the Suds-i Fakhri of al-Khujandi, it was a venerable instrument in Islamic astronomy. At Samarqand it was Ulugh Beg's largest and most visible construction. The Shahjanabad sextant had a radius of 8.25 metres and seems to have been extremely accurate. It yielded readings of one minute of arc or better – the limit of the naked eye.[25] The Daksinottara Bhitti Yantra was a mural quadrant. Located on a north-south wall, it was used to measure the meridian altitude or zenith distance of a heavenly object – Sun, Moon, planet, or star. Ptolemy described a portable version of the instrument, and Ulugh Beg had constructed a large mural quadrant in Samarqand. In 1729, Jagannatha used the mural quadrant in the Shahjahanabad Jantar Mantar to determine the obliquity of the ecliptic.[26]

The Jaya Prakash consisted of two hemispheric masonry bowls 8.33 metres in diameter. It was a multipurpose instrument designed to determine several astronomical parameters. Its graduated arcs measured the azimuth angle and zenith distance of celestial objects. It also indicated the signs of the zodiac.[27] The Rama Yantra was a cylindrical structure with two complementary halves; a pillar was set in the middle of each half. It was used to measure the azimuth and altitude of a celestial object. The instruments in Shahjahanabad were twice as large as those in Jaipur: the cylinders were 16.65 metres in diameter and the pillars were 7.52 metres high.[28] The last instrument constructed by Jai Singh in the capital was the Agra or sundial. It was placed on the top of the Samrat gnomon and was 1.5 metres high and 1.5 metres in diameter.[29]

The Shahjahanabad Jantar Mantar was the first and most important of Jai Singh's five observatories. It was the observatory of the *Zij-i Jadid-i Muhammad Shahi* – the place where observations were made, tables compiled, and the text drafted. Its precedence reflected Muhammad Shah's role: his approval launched the project and Shahjahanabad was the Mughal capital. On the other hand, the Jantar Mantar in Jaipur, the Amber capital, contained more instruments – thirty (some are replicas) rather than six – and employed more astronomers for a longer period. Construction probably began in 1723–4, several years after the Shahjahanbad observatory, and continued through to 1738. The two most important instruments (the Samrat and Sasthamsa Yantra) were erected between 1732 and 1735.[30] Jai Singh had been governor of Malwa, whose capital was Ujjain. For centuries the centre of Indic astronomy, Ujjain was the site of the prime meridian for ancient India, and the Raja's decision to construct a small observatory there was a reflection of its status. Built between 1724 and 1730, the Ujjain Observatory contained only six instruments and ceased to operate after Jai Singh's death in 1743.[31] The Varanasi Observatory was assembled on the roof of a tall building overlooking the Ganges. Probably constructed between 1724 and 1730, it boasted only seven instruments – none as large or as accurate as the ones in Shahjahanabad or Jaipur.[32] Jai Singh also laid out an observatory in the north Indian city of Mathura. Probably erected between 1724 and 1730, it was located on the roof of a building in a local fort and contained only six small instruments. Not much is known of its operation and it seems to have been the least important of the five.[33]

Like Nasir al-Din Tusi, Ulugh Beg, and Taqi al-Din, Jai Singh recruited a substantial stable of astronomers. Because his early education was in Indic mathematics and astronomy, the great majority of his astronomers were Indic. His chief lieutenant was Jagannatha Samrat, whom he convinced to learn Persian and Arabic and to master the Islamic classics – two of which he translated into Sanskrit. *Samrat Siddhanta* was a translation of Nasir al-Din Tusi's recension of the *Almagest*, and *Rekhaganita* was a translation of Tusi's recension of Euclid's *Elements*. Jagannath also wrote *Yantraprakara*, a work on astronomical instruments that reproduced some of the material from his *Samrat Siddhanta*. The construction and operation of the Shahjahanabad Jantar Mantar was probably under his supervision as well.[34] Besides Jagannath, Jai Singh employed many other Indic astronomers – a 1735 list of the men at the Jaipur Jantar Mantar contained twenty-two names. They erected instruments and took measurements, authored texts and commentaries, translated works from Arabic and Persian into Sanskrit, and copied books for the Raja's library.[35]

In addition to the Indic astronomers, Jai Singh patronised Muslim and European astronomers as well. Dayanat Khan seems to have been Jai Singh's favourite Muslim astronomer, working for the Raja from 1718 to 1739, but the Amber state archives listed at least eleven others as well. These men built the early brass instruments and were heavily involved in the construction of the Shahjahanabad and Jaipur Jantar Mantars. They also translated the Persian and Arabic treatises and assisted in the drafting of the *Zij-i Jadid-i Muhammad Shahi*.[36]

The smallest group of astronomers in Jai Singh's employ (and the last to be recruited) were the Europeans – Portuguese Jesuits. By the time of Muhammad Shah's accession, the Portuguese had been in India for more than 200 years. Controlling a 60-mile wide swath of territory south of Bombay, the Portuguese had scattered throughout north-western India as mercenaries, traders, and adventurers. The Jesuits, sent out to minister to their countrymen and to convert the populace, included astronomers and mathematicians among their number. The attraction of the two parties – Jai Singh and his Indic and Muslim astronomers, on the one hand, and the Jesuit scientists, on the other – was mutual. Jai Singh wanted to learn more about European instruments and theories and the Jesuits wanted to secure a place of influence at the Raja's court. Because Jai Singh's 1720 petition to Muhammad Shah mentioned European astronomical tables, the first appearance of the Portuguese astronomers was probably in the second decade of the eighteenth century – between 1714 and 1719 perhaps.

After constructing the Shahjahanabad Jantar Mantar and gathering several years of data, Jai Singh decided to find out more about European astronomy. In the preface to the *Zij-i Jadid-i Muhammad Shahi* we read: 'After seven years had been spent in this effort (observing the stars), information was received that observatories had been built in Europe . . . and that the business of the observatory was still being carried on there.'[37] In 1727, therefore, the Raja decided to organise and fund a scientific mission. Headed by Father Manuel de Figueredo, the Rector of the Jesuit College in Agra, its destination was Lisbon and it included both a Muslim and a Christian (convert) astronomer. The delegation finally arrived in January 1729, and on 10 March the *Gazeta de Lisboa* reported that the Indians had come to find out more about European astronomical instruments and to resolve questions regarding the astronomical tables of Portugal and India. The group remained in Portugal (mostly in Lisbon) for about a year. They do not seem to have visited either Paris or London – the chief centres of European astronomy. At least there was no mention in the local papers of a delegation to either the Paris Observatory (finished in 1671) or the Royal Greenwich Observatory (finished in 1675). The delegation returned to Jaipur in July 1731. They brought back instruments, books, and tables – in particular the *Tabulae Astronomicae* of Philippus de La Hire (1640–1718), the French mathematician and astronomer. His was the latest and most accurate of the European treatises.[38] In the following decade the names of several Portuguese astronomers appeared in the records of the Amber state: Pedro de Silva, a Lisbon physician and amateur astronomer, returned with the delegation to India; Father Du Bois translated part of de La Hire's work; and Father Boudier answered the Raja's queries about planetary motion. In addition, Jai Singh invited two eminent Jesuit mathematicians to come to Jaipur to consult on mathematical astronomy. The Jesuit astronomers also helped Jai Singh purchase European instruments and acquire and translate European books and treatises.[39]

In the Shahjahanabad and Jaipur Jantar Mantars, as in the observatories at Maragha and Samarqand, an impressive astronomical library – works on astronomy, astrology, and mathematics – aided the work of the resident scientists. At the time of Jai Singh's accession to the Amber throne, the royal library contained only thirty-two such works. By the early 1730s, however, when the *Zij-i Jadid-i Muhammad Shahi* was being compiled, the Raja's library had swelled to more than two hundred titles. Beyond the expected works in Sanskrit, Persian, and Arabic, there were books and treatises in Hindi, Bengali, Latin, French, German, and English. Some of these, the European especially, were very expensive while others had been borrowed from the imperial Mughal library or had been copied from the holdings of private individuals. The materials were kept in Jai Singh's personal library or were distributed among the observatories – primarily those in Shahjanabad and Jaipur. The Raja's library also included astronomical instruments: twenty-four astrolabes, several celestial globes, a set of European drawing instruments, and a simple telescope.[40]

For Jai Singh, as for Nasir al-Din Tusi and Ulugh Beg, the motivation behind recruiting astronomers, constructing observatories, and stocking libraries was the compilation of an accurate set of astronomical tables – so as to correctly predict the motions of the seven heavenly bodies and the eclipses of the Sun and Moon. While the result of all this work – *the Zij-i Jadid-i Muhammad Shahi* – was more accurate than the *Zij-i Ilkhani* or the *Zij al-Sultani*, it was curiously old-fashioned, a medieval geocentric exercise in an early-modern heliocentric world.

Observations and data gathering began at the Shahjahanabad Jantar

Mantar in 1721, soon after the first instruments were erected, and continued until 1731–2, after the return of the European mission. Jai Singh's treatise took the *Zij al-Sultani* of Ulugh Beg, finished in c. 1440, as its model. Like the Samarqand work, it was divided into three chapters: Chapter One on calendars, Chapter Two on trigonometry and spherical astronomy, and Chapter Three on planetary motions. It, however, lacked a fourth chapter on astrology.[41]

Chapter One contained only four sections. The first covered the Indic Vikramaditya Samrat (Vikramaditya Era), named after King Vikramaditya of Ujjain, whose epoch was 14 October 58 BCE. The second was the Muslim Hijra Era and the third was the Muhammad Shahi Era. Named after the reigning monarch, its epoch was 20 February 1719 – the date of Muhammad Shah's accession. The fourth section laid out the rules for converting dates from one calendar and era to another.[42]

Chapter Two of the *Zij-i Jadid-i Muhammad Shahi*, like the corresponding chapter of the *Zij al-Sultani*, dealt with trigonometry and spherical astronomy. It contained nineteen sections and tables and included daylight equations for Jaipur and Shahjahanabad and the latitudes and longitudes of selected cities, the longitudes measured from the Shahjahanabad Jantar Mantar and expressed in Indic measures of time. Section four reported the obliquity of the ecliptic at different places: Samarqand, cities in Europe, and all five of Jai Singh's observatories.[43]

Chapter Three on planetary motion contained thirteen chapters (with tables). The epoch for the tables was that of the Muhammad Shahi Era (20 February 1719). The planetary tables included the main elements for the twelve months of the Hijra year at monthly intervals, for thirty Hijra years at yearly intervals, and for 390 Hijra years at thirty year intervals. The tables for the Moon listed the Moon's first, second, and third equations. The chapter also detailed procedures for calculating eclipses of the Sun and Moon. Two examples for Jaipur were given: a lunar eclipse on 8 June 1732 and a solar eclipse on 2 April 1734. The times calculated were accurate to within three minutes. The conclusion to Chapter Three included star tables, the constellations arranged in the same order as in the *Zij al-Sultani*. The epoch of the tables was 1725–6, and Jai Singh's astronomers seem to have made new observations of about sixty stars. However, the positions of the remaining stars in Ulugh Beg's list (a total of 1018) were obtained by applying a precession correction.[44]

In the end, however, given the advances in European science – the telescope and the theories of Copernicus (1473–1543), Kepler (1571–1630), and Newton (1642–1726) – the astronomical work of Jai Singh – both observatories and treatises – remained completely medieval – closer to Ptolemy than to Copernicus. The Raja was aware of the telescope. His library contained a simple, India-made example which he had purchased for Rs. 100 but he chose not to employ it in his observational work.[45] Of the instrument he wrote:

These rules are for naked eye observations only, although the telescope is now being made in the country. The telescope enables one to see bright stars in broad daylight also – say around the noon hour. It also enables one to see the moon when there is hardly any light in it, or when its face is totally dark and invisible . . . The planet Saturn (through a telescope) appears oval in shape, an oval whose lower half is larger than the upper. Around the planet Jupiter there revolve four bright stars. On the face of the sun there are spots, and the sun rotates once on its axis with a period of one Year . . . Since the telescope is not readily available to an average person, we are going to base our rules of computations for the naked eye only.[46]

The Raja's decision was not the result of a blind adherence to tradition. Rather, until the invention of the telescopic sight in 1667, the simple telescope didn't provide a great advantage over the naked eye and even after its first appearance it was several decades before the improved sight offered significant advantages. As for the heliocentric theories of Copernicus and the work of Galileo and Kepler, they were unknown to the Raja.

Nicolaus Coperrnicus published *De Revolutiones* in 1543. As his heliocentric argument became gradually more accepted, the Catholic Church mounted an extensive and increasingly violent campaign of opposition. In 1600 Giorduno Bruno, the Dominican mathematician and astronomer, was arrested and burned alive for proclaiming that the Sun was at the centre of the heavens. And in 1616 the church condemned the views of the Italian mathematician and astronomer Galileo Galilei. In 1633 Galileo was brought before the Inquisition and convicted of heresy. Forced to recant, he was sentenced to house arrest for the last ten years of his life. Given this climate it is no surprise to find that among Catholic scholars and intellectuals the theories of Ptolemy reigned supreme. For example, an encyclopaedia of mathematics published by the Jesuits in China in 1645 was Ptolemaic, and a 1738 astronomical treatise by another Chinese Jesuit was also geocentric.[47] Thus, it is perfectly understandable that Jai Singh's Jesuit astronomers should have brought him no news of Copernicus, Kepler, Galileo, and Newton. As for de La Hire's *Tabulae Astronomicae*, first published in 1687, with a new edition in 1702, Jai Singh had copies of both. However, he made almost no use of the work in his treatise. Since he had already adopted the format of the *Zij al-Sultani* and since de La Hire's treatise was for a European audience, his only appropriation from the *Tabulae Astronomicae* was its refraction-correction tables. Islamic treatises had never contained such information.[48]

Notes

1. Lapidus, *Islamic Societies*, 437–67.
2. B. V. Subbarayappa (ed.), 'The Tradition of Astronomy in India, Jyotihsastra'; S. A. Ghori, *Appendix III: Development of Zijes in India* (New Delhi: Center for Studies in Civilisations, 2008), 393.

3. Ibid. S. M. Razaullah Ansari, *Appendix 1: On the transmission of Islamic Astronomy to Medieval India*, 391.
4. Ghori, *Development of Zijes in India*, 351–2.
5. Ansari, *On the transmission of Islamic Astronomy*, 394–5.
6. Sreeramula Rajeswara Sarma, *Astronomical Instruments in the Rampur Raza Library* (Rampur: Rampur Raza Library, 2003), 13.
7. Ghori, *Development of Zijes in India*, 391–2.
8. Sarma, *Astronomical Instruments*, 7–8.
9. Ansari, *On the Transmission of Islamic Astronomy*, 391–2.
10. Ibid. 395–6.
11. Blake, *Time*, 122–6.
12. Ibid. 118–40.
13. Ibid. 130–1.
14. Ibid. 59–63; Pingree, *Omens to Astrology*, ch. 7.
15. Blake, *Time*, 60–1; Pingree, *Omens to Astrology*, ch. 7.
16. Ghori, *Development of Zijes in India*, 392.
17. Ibid. 396.
18. Ibid. 386.
19. Ibid. 396–7.
20. Nirendra Nath Sharma, *Sawai Jai Singh and his Astronomy* (Delhi: Motilal Banarsidass Publishers, 1995), 2.
21. Ibid. 2–4.
22. Ibid. 19–20.
23. Ibid. 24, 36–8.
24. Ibid. 41–57, 99–103.
25. Ibid. 58–62; 103–7.
26. Ibid. 63–6.
27. Ibid. 67–72, 106–7.
28. Ibid. 80–3; 107–8.
29. Ibid. 119.
30. Ibid. 121–90.
31. Ibid. 212–28.
32. Ibid. 190–211.
33. Ibid. 229–33.
34. Ibid. 257–64.
35. Ibid. 265–81.
36. Ibid. 283–6.
37. Ibid. 295.
38. Ibid. 245.
39. Ibid. 287–303.
40. Ibid. 253–6, 332.
41. Ibid. 234–52.
42. Ibid. 235–6.
43. Ibid. 236.
44. Ibid. 241.
45. Ibid. 243.
46. Ibid. 235–36.
47. Ibid. 291–2, 311.
48. Ibid. 245–50.

CHAPTER NINE

Medieval and early-modern Europe

Just as in the Muslim world the interest in astronomy was tied to the require-
ments of the faith – times of prayer, appearance of the new Moon, orientation
of the mosque – so too in the Christian. By the tenth century the Latin West
had two well established traditions of practical astronomy. Ecclesiastical *computus* (*calculation*) required rudimentary arithmetical calculations to determine
the dates of Easter and its related feasts. Furthermore, simple observations of
the Sun and stars enabled monastic communities to determine the times for
both daytime and night-time prayers.[1]

By the middle of the first millennium one of the most important issues for the
early church was the date of Easter. The Christian calendar featured both solar
and lunar festivals. Christmas on 25 December – the medieval date of the winter
solstice (astronomically December 21) – was solar. Easter, on the other hand, tied
to the Jewish Passover, was lunar. Jesus of Nazareth and his early followers were
Jews, and his execution occurred in Jerusalem during the Passover celebration.
The Last Supper (the final communal meal for Jesus and his followers) was a
Passover feast (on Maundy Thursday). Jesus was crucified on Friday and, according to the early accounts, rose from the grave on Sunday. To retain the New
Testament chronology, therefore, Easter had to be celebrated after Passover. In
the lunisolar calendar of the Jews, Passover occurred on the first full moon after
the vernal equinox (21 March). Thus the early church decided that Easter should
be celebrated on the first Sunday after the first day of Passover. For the medieval
scholar, calculating the date of Easter was a complicated calendrical (and astronomical) problem. Determining the exact moment of the equinox and integrating
the two years – the 365 ¼ day solar and the 355 day lunar – was extraordinarily difficult. So important, in fact, was this calculation that it acquired its own
name – the *computus* (calculation). In 525 CE Dionysius Exiguous (c. 470–545),
a Romanian monk, published a table establishing the dates of Easter for the
next ninety-five years but at that time the Christian calendar employed the Era
of Diocletian. Named after an emperor who had persecuted the early church,
the era was solar – its epoch 1 January 284. Because Dionysius did not want to
perpetuate the memory of a tyrant, he launched a new era: *Years of Our Lord Jesus
Christ, Anni Domini Nostri Jesu Christi* (Anno Domini or AD). Although the Venerable
Bede in *The Ecclesiastical History of the English People* (731) used Dionysius' era, the
Roman Church did not fully adopt it until the early tenth century.[2]

Besides the calendar, the other issue for medieval Europeans was timekeeping. For the ordinary person, the division of the day was relatively simple: daily prayers in the morning (Lauds) and evening (Vespers) and a communal service (time unspecified) on Sunday. In the monasteries, however, a more structured arrangement was adopted. The Benedictine Rule, introduced by St Benedict in 540 and modelled after the Roman army, organised the monks' day around eight prayers: Matins (early morning, still dark), Lauds (very early morning), Prime (first hour), Terce (third hour), Sext (sixth hour), None (ninth hour), Vespers (when the evening star first appears), and Compline (just before retiring).[3]

Until the encounter in the twelfth century with the Latin translations of the Greek and Islamic astronomers, European astronomy was taught as one of the seven liberal arts: the first three – grammar, rhetoric, and dialectic – and the more advanced four – geometry, arithmetic, astronomy, and music. Medieval European astronomy was descriptive and philosophical, based on Aristotle's *On the Heavens*. Finished in 350 BC, this treatise provided, as we have seen, a description of the heavenly world and the ways in which it differed from the terrestrial. According to Aristotle, the Earth was composed of four perishable elements whose motion was linear. The heavenly bodies, on the other hand, were composed of imperishable ether, and their motions were circular. Aristotle's cosmology remained a staple of the university curriculum for centuries.[4]

The initial encounter of European scholars with the work of the Greek and Arabic astronomers was in Islamic Spain. In the Umayyad capital of Cordoba, al-Majriti, 'Abd al-Rahman III's court astrologer, put together an adaptation of al-Khwarizmi's *Zij al-Sindhind*. In addition to introducing the Europeans to the Islamic *zij* in its standard Ptolemaic format (chronology, motions of the Sun and Moon, eclipses, planetary equations, star catalogue), the *Zij al-Sindhind* also included Khwarizmi's work on algebra and Arabic (or Hindu) numbers. After al-Majriti, the most important early Andalusian astronomer was al-Zarqali, head of the Toledo school of astronomers. Al-Zarqali was a talented instrument maker as well as a gifted astronomer. He invented a universal astrolabe (called in Europe the Saphea Arzachelis or Disc of Zarqali) but his major claim to fame, in both Spain and northern Europe, was the *Toledan Tables*. Taking al-Khwarizmi as his starting point, al-Zarqali calculated a new value for the solar apogee, updated the star catalogue, and adopted al-Battani's solar, lunar, and planetary equations.

The new learning began to spread to northern Europe in the late tenth century, accelerating rapidly through the eleventh and twelfth centuries. Gerard of Aurillac (d. 1003), later Pope Sylvester II (999–1003), was one of the earliest transmitters. His contribution was more the proclamation of a whole new body of knowledge than it was the passing on of specific facts or theories. He was followed by a number of other early scholars – French, English, and Italian.

Adelard of Bath (c. 1080–1150) produced the first translation of al-Khwarizmi's *zij* (c. 1134) and composed a work on the astrolabe. He also translated Euclid's *Elements* and many Arabic works on astrology and magic. Prior Walcher of Malvern Abbey (1091–1125) in England, read Adelard's translation of al-Khwarizmi, and began to make his own observations. Using an astrolabe (perhaps of al-Zarqali's design) he predicted the time of several solar and lunar eclipses and computed a set of new Moon tables for the period 1036–1111.[5] The Italian Gerard of Cremona (c. 1114–87) travelled to Toledo, learned Arabic, and translated eighty-seven works of Greek and Islamic astronomy and astrology from Arabic into Latin. Among these were Aristotle's *On the Heavens*, Euclid's *Elements*, al-Zarqali's *Toledan Tables*, and, most famous of all, Ptolemy's *Almagest* (1175). As the *Zij al-Sindhind* and the *Toledan Tables* became more widely known, they were adapted by local scholars for the latitudes of London, Paris, and Pisa.[6]

In addition to the treatises, the other Islamic influence on European astronomy and astrology in the eleventh and twelfth centuries was the astrolabe. By c. 1030, texts (translated from Arabic into Latin) and working models had become widely available. A versatile and valued instrument, the astrolabe was as much a practical timekeeper as it was an observational instrument. In the late thirteenth century European astronomers added a thread which allowed the measurement of planetary altitudes, turning the simple astrolabe into a new instrument – the astrolabe-quadrant.[7]

The last astronomical treatise from Islamic Spain was completed in Toledo at the court of Alfonso X (1252–84), Christian ruler of Castile, Leon, and Galicia. Alfonso had established a school of Christian and Jewish translators who turned the Arabic works of astronomy and astrology into Castilian and Latin. But like Malik Shah and Hulagu, he also wanted a revision of the outdated *zij* of his day – the *Toledan Tables* of al-Zarqali. He ordered his astronomers to construct new equipment, make fresh observations, and prepare a new set of tables. The *Alfonsine Tables* (finished in c. 1272) introduced the Alfonsine Era, epoch 1252. As in the earlier *zij*s, rules were provided for converting dates from the Alfonsine Era into the Hijra and Yazdegird eras and vice versa.[8]

In the history of European astronomy and astrology, however, the *Alfonsine Tables* were an important milestone. In 1321 the Parisian astronomer John of Murs composed a treatise entitled *Exposition of the Meaning of King Alfonso in Regard to His Tables*. Several years later John of Ligneres, teacher of John of Murs, composed canons for a new edition of the tables. This edition included the twelve signs of zodiac in the Islamic configuration, an improved set of planetary equations, and new values for the solar eccentricity. It also contained the Alfonsine theory of precession, the Spanish value for the solar equation, and an Andalusian star catalogue.[9] As we have seen, it seems likely that the Andalusian *zij* was, at the very least, an important stimulus for the early European astronomers, whose subsequent versions of the *Alfonsine Tables* dominated the European scene for the next two centuries.

In the early thirteenth century, universities were founded in Oxford, Bologna, and Paris. The astronomy they taught was primarily descriptive (Aristotlean), and included none of the mathematics and geometry of Ptolemy and the Islamic astronomers. As the Latin translations of the Arabic works became more widely available, however, the universities began to take notice, and a new genre of books called *theorica planetarum* (theory of the planets) began to appear. One of the best known was John of Seville's *Theory of the Planets* (1135), a translation of al-Farghani's *Elements of Astronomy on the Celestial Motions*. This treatise, a descriptive, non-mathematical summary of Ptolemy's *Almagest*, gave the student the essential elements of solar, lunar, and planetary models without the rules and lists of the ordinary *zij*.[10] The Parisian astronomers of the late-thirteenth century also took an interest in astronomical instruments. They began to use mural quadrants and parallactic rulers. John of Ligneres, for example, composed treatises on the armillary sphere, the universal astrolabe (after al-Zarqali), the equatorium, and the directorium (a calculating instrument related to the astrolabe).[11]

Like the University of Paris, Oxford University during the early-thirteenth century taught Aristotle and the *Toledan Tables*. By the early fourteenth century, however, Oxford could boast one of the most talented astronomers of the medieval period. Richard of Wallingdon (c. 1292–1336), a Benedictine monk, was first an Oxford don and later abbot of St Albans monastery. He composed the first comprehensive treatise on spherical trigonometry in northern Europe, the *Quadripartitum*. Based on the *Almagest* and the *Toledan Canons*, Richard's treatise offered exact solutions to problems in spherical geometry. The calculations, however, were tedious, and Richard invented a new instrument (the rectangulus), to simplify the process. He also wrote *Treatise on the Albion* (a planetary equatorium). Although the instrument was difficult to build, it was extremely versatile. No problem in classical astronomy was beyond it: calculating parallaxes, velocities, conjunctions, oppositions, and eclipses of the Sun and Moon. He constructed a sophisticated mechanical clock at St Albans, featuring an escapement and displaying lunar phases, eclipses, planetary movements, and tidal schedules.[12]

Although Richard finished his work on the *Albion* in 1327, he made no mention of the *Alfonsine Tables*. In 1330 a Parisian edition of the tables arrived in Oxford, and in 1340 a university astronomer created a new version with decimal units instead of sexagesimal, an Oxford meridian replacing the one of Paris. In 1348 another Oxford edition appeared with double entry tables giving planetary movements and positions. Of great use for astrologers, this version spread rapidly throughout Europe. In the early fifteenth century the Oxford astronomer John Killingworth (c. 1410–45) produced a revised edition of the *Alfonsine Tables*, from which a full ephemeris could be created.[13]

By the middle of the fifteenth century the number of European universities had greatly increased, and the passion for astronomy had developed apace – especially in the German-speaking countries. Two of the most important northern astronomers were Georg Puerbach (1423–61) and Regiomontanus (1436–76). Puerbach was an Austrian who served as court astrologer to the Hungarian ruler, Ladislaus V (1444–57). He wrote *Theoricae novae planetarum* (*New Theory of the Planets*) and taught Regiomontanus at the University of Vienna. Printed by Regiomantanus in 1474, Puerbach's *Theoricae* depended heavily on the 1348 Oxford edition of the *Alfonsine Tables*. And it was immensely popular, going through nearly sixty editions before being superseded in the early seventeenth century. In the late 1450s Puerbach compiled *Table of Eclipses*, and several years later he began a new abridgment of Gerard of Cremona's translation of the *Almagest*. After his death, the work was completed by Regiomontanus.

In 1471 Regiomontanus, having become increasingly famous as an astronomer and astrologer, moved to Nuremberg. The leading commercial centre of central Europe, Nuremberg offered precision instruments and state-of-the-art printing presses. After publishing Puerbach's *Theorica* in 1474, Regiomontanus printed his own *Kalendarium and Ephermerides*, an astronomical calendar and almanac (in both Latin and German for the years 1475–1506). Columbus is said to have had a copy of this work on his fourth voyage and to have astonished the Jamaican Indians by predicting the lunar eclipse of 29 February 1504. Regiomontanus's *On Triangles*, first published in 1533, made use of the so-called cosine law as well as the sine law for spherical triangles.[14]

Nicolaus Copernicus (1473–1543) marked the beginning of the end for the superiority of Islamic astronomy and astrology. Although the theorems of several Islamic astronomers appeared in his heliocentric system, by the end of the sixteenth century leadership in the science had passed from the Islamic world to the European. Copernicus was born in Poland, the subject of a Polish King, but he was a German humanist at heart, heavily influenced by German culture, the Catholic Church, and Renaissance Italy. He studied law and Aristotelian astronomy at the University of Krakow, his textbook Peurbach's *New Theory of Planetary Motion*. In 1493 his uncle named him canon (legal official) of the Cathedral of Varmia in north-eastern Poland. In 1496 he took a leave of absence so that he could attend the University of Bologna where he studied canon law, Greek, and astrology, reading Ptolemy in the original. On returning to Poland he resumed his churchly duties. Since his post offered a generous lifetime income, he remained in Varmia for the remaining forty years of his life. His passion for astronomy led him to build an observation tower with three instruments: an armillary, a quadrant, and Ptolemy's parallactic ruler. In 1510 he completed *Commentariolus* (*Brief Commentary*). In this treatise he presented his basic hypotheses without offering any argument or evidence. During the years 1512–19 he completed a number of observations, and in 1543, the year of his death, he finally published *De revolutionibus orbium caelestium* (*On the Revolutions of the Celestial Spheres*). Because he was worried about his work's reception, he included an anonymous preface by a Lutheran theologian, Andreas Osiander. Osiander had advised Copernicus that astronomical theories were not articles of faith. They were simply convenient bases for computation. As long as they yielded accurate results, their truth or falsity was of no importance. Although Copernicus did not agree with this point of view – his system, he believed, represented physical reality – many early readers took it to represent his opinion.

In summary, Copernicus's contributions in *De revolutionibus* were:

1. he explained the diurnal rotation of the celestial sphere as a consequence of the axial rotation of the Earth;
2. he explained the Sun's apparent motion in the ecliptic and the stations and retrograde motions of the planets as a consequence of the Earth's annual revolution around the Sun;
3. he explained planetary motions from a heliocentric point of view;
4. he explained the retrograde motions of the planets;
5. he determined the relative distances of the planets from the Sun;
6. he adopted a system of eccentric circular deferents for the planets, thereby decreasing the number of circles to five, one for each planet.

Copernicus's system, though revolutionary in implication, was in formulation strangely conservative. In many of his arguments and assumptions, his work remained within a Ptolemaic framework. Although the Sun replaced the Earth in Copernicus's model, in most other respects he followed the planetary model of Ibn al-Shatir. He retained the Aristotelian principle of uniform circular motion and kept Ptolemy's system of eccentric epicycles, employing the 'Tusi Couple' to rid his model of the infamous equant. Like Ptolemy's *Almagest*, his *De revolutionibus* was divided into six books. The first gave a general survey of his heliocentric system. Copernicus did not merely replace the Earth with the Sun but he also established a definite order for the planets. The Sun was at the centre followed by Mercury, Venus, Earth (circled by the Moon), Mars, Jupiter, and Saturn. He established their rotational periods, explained the shape of their retrograde motions, and approximated their relative distances from one another. The second book offered a basic explanation of spherical astronomy. The third concerned precession and the motion of the Earth. The fourth dealt with the Moon. Copernicus's model was almost identical to that of Ibn al-Shatir, and his parameters fit those of the *Alfonsine Tables*. His discussion of the distance, parallax, and apparent diameter of the Sun and Moon were better than Ptolemy's but still in need of revision. He was much better, however, at calculating eclipses. The fifth book took up planetary longitude. Although Copernicus rewrote the Ptolemaic epicycles for the superior planets, the centre of his system was not the true Sun but the centre of the Earth's rotation around the Sun. His work on the inferior planets was less valuable because he could not obtain the observations he needed. His sixth book on planetary latitudes was also flawed because the orbits of his planets did not centre on the true Sun.

As a general principle, Copernicus seems to have wanted to retain as much of his predecessors' work as he could. His system, like Ptolemy's, was unified but geometrical. It was not yet a physical system, explaining the movements of the heavenly bodies in terms of physical laws. Kepler remarked that Copernicus did not realise how rich his model was. He put more effort into interpreting Ptolemy than into deciphering nature.[15]

Just as the medieval Europeans received inspiration from Islamic astronomers, so too the scholars of Renaissance Europe – even so modern and radical a thinker as Nicholas Copernicus. Although it is undeniable that his central insight – that the Earth circled the Sun while revolving on its axis – was revolutionary, recent research has shown that Copernicus, in addition to demonstrating a great deal of loyalty to Ptolemy, also utilised the work of later Islamic astronomers. His lunar theory followed almost exactly that of Ibn al-Shatir, who had employed in his planetary equations both the 'Tusi Couple' and the ''Urdi Lemma'. As a result, Copernicus was able to rid his models of the Ptolemaic equant and to reduce the number of epicycles, allowing a more accurate set of predictions. Copernicus's reliance on the Ptolemaic framework as well as his use of Tusi and 'Urdi caused two scholars in a recent book on Coperrnicus's mathematical astronomy to label him the last of the Maragha astronomers.[16]

The remaining puzzle, however, is the path of transmission, the mechanism by which the theories of these thirteenth-century Islamic astronomers reached the eyes of a sixteenth-century Polish astronomer. On the one hand, the earlier pathway was no longer available. The centre of Islamic astronomy had moved eastward, from Baghdad, Damascus, and Cairo to Margaha and Samarqand, and the Andalusian centres of Cordoba, Toledo, and Seville had returned to the Christian fold. The Latin translations of the eleventh–thirteen centuries had jump-started medieval European astronomers, who had begun their own programmes of observation, calculation, and theorising. The *Zij-i Ilkhani* (1271–2) of Nasir al Din Tusi, almost exactly contemporaneous with the *Alfonsine Tables* (1272) of Alfonso X, but more accurate and mathematically superior, and the *Zij al-Sultani* (1437) of Ulugh Beg, more advanced again, were both completely unknown in Renaissance Europe.

The direction of transmission was new – from the East rather than from the South – and the language of translation was also new – from Greek rather than from Latin. When Constantinople fell to the Mehmed II in 1453, the Byzantine astronomers fled to the early-modern West. For Renaissance humanists, eager to recover their ancient heritage in the original, knowledge of Greek had become de rigueur. As a result, Copernicus is thought to have read the commentary of a Greek astronomer on Ibn al-Shatir, where he would have encountered diagrams and mathematical proofs of the 'Tusi Couple' and the "Urdi Lemma'. In addition, by the early sixteenth century European scholars had learned Arabic and had begun to travel to Istanbul. Copernicus could well have acquired knowledge of the Maragha astronomers from men like Guillaume Postel (1510–81), the French linguist and mathematician who spent several years in Istanbul and other Eastern Islamic cities. He brought back the astronomical and mathematical manuscripts of Tusi and his collaborators. In the transmission of scientific knowledge from the Islamic East to the European West the situation had changed dramatically from the thirteenth century to the sixteenth. An intermediary (translator or importer) was no longer needed; the connection was direct. Andreas Vesalius (1514–64), the founder of modern anatomy, revealed the extent of that change when he wrote '. . . those Arabs are now rightly as familiar to us as are the Greeks.'[17]

Tycho Brahe (1546–1601) and Johannes Kepler (1571–1630) bridged the gap between Nicholas Copernicus and Isaac Newton (1643–1726). Each provided theories and observations that led from the heliocentric but geometric model of the first to the gravitational, physical model of the second. Brahe, born into an aristocratic family in Denmark, entered the Lutheran University of Copenhagen at age thirteen. A partial eclipse of the Sun in 1560 sparked a nascent interest in astronomy. Between that eclipse and the Jupiter–Saturn conjunction of 1563, he made enough observations with his primitive cross-staff to demonstrate major errors in the two astronomical treatises of his day: up to five degrees in the *Alphonsine Tables* and up to four degrees in the *Prutenic Tables* (1551). Over the next ten years Brahe travelled extensively in Germany, Switzerland, and Italy, attending the universities of Leipzig, Wittenburg, Augsburg, and Basel. Although he concentrated mostly on mathematics and physics, he also became an expert alchemist, astrologer, and chemist. On 11 November 1572 Brahe observed a new star (a supernova) in the constellation Cassiopeia. It was brighter than Venus and located, he argued, in the distant stellar regions beyond the planets, not in the sublunar sphere between the Earth and the Moon. He sent his 1573 book, *De nova stella* (*On The New Star*), to a number of scientists and by 1575 his fame had spread throughout Europe. The comet of 1577, the one that played such an important role in the Istanbul Observatory of Taqi al-Din, offered him another opportunity to display his skills. For three months he kept it under careful observation. At first he located it above the sphere of the Moon, later within the sphere of Venus, before finally deciding that it was even farther out, beyond the sphere of the Sun.

In the spring of 1575 Brahe visited William IV (1532–92), Landgrave of Hesse-Kassell, and spent a week working in the nobleman's small observatory. In 1576 King Frederick II (1534–88) of Denmark, another amateur astronomer, offered Brahe the island of Hven in the Danish Sound and a lavish grant (5 per cent of the Danish GNP) to build and furnish a new observatory. Uraniborg

(Castle of Urania, the muse of astronomy) was the most advanced of its day, and Brahe worked there for twenty years. The observatory boasted the most complete and advanced collection of astronomical instruments in the pre-telescopic era: Ptolemaic rulers, armillaries, sextants, octants, azimuthal quadrants (of wood and of brass), and celestial globes (the largest a metre and a half across). Uraniborg's finest instrument, however, was a mural quadrant with a radius of 1.8 metres. It required three assistants: one to observe the object through the pinnules on the sighting rule, one to enter the results in a ledger, and a third to note the time on two nearby clocks. Uraniborg also included a windmill, paper mill, printing press, bindery, forge, chemical ovens, laboratories, pharmaceutical gardens, and a zoo. In 1584 another structure – Stjerneborn (Castle of the Stars) – was added. It contained instruments in subterranean rooms mounted on secure foundations,

In 1588, Tycho finally published his study of the 1577 comet – *De mundi aetherii recentioribus phaenomenis* (*Concerning Recent Phenomena of the Aetherial World*) – and sent it to a number of scientists including Galileo. By then his doubts about the Aristotelian spheres had crystallised. Since they did not seem to impede the motion of a comet beyond the Moon or the birth and death of a star, they could not be solid or real in any sense. Although earlier Islamic astronomers had questioned the Aristotelian doctrine, Brahe is generally considered to have laid it finally to rest. The implications were sweeping. If the planets were fixed on hard, material spheres, their movements were simply a consequence of the movements of the spheres themselves. Without the spheres, however, a new theory of planetary motion was needed. What were the real dynamics of the solar system?

In 1588 also, Frederick II died and was succeeded by his eleven-year-old son, Christian IV (1588–1648). Demanding and quarrelsome, Tycho soon fell out with the new regime, and in 1597 he left Hven for Hamburg. In 1598 he published *Astronomiae instaurantae mechanica* (*Instruments of the Restored Astronomy*). An illustrated account of the buildings and instruments at Uraniborg, he dedicated the book to the Holy Roman Emperor Rudolf II (1575–1612). In 1599 he accepted the patronage of Rudolph and moved to Prague, and in 1601 he died. In Prague he met and hired Johannes Kepler, who oversaw the final publication (1602) of his great work – *Astronomiae instauratae progymnasmata* (*First Exercises in a Restored Astronomy*).[18]

Brahe's greatest contribution may well have been his insistence on an extended and continuous series of observations of the Sun, Moon, and planets – rather than a small number of separate sightings on significant occasions. In his preface to the *Rudolphine Tables* (begun by Brahe) Kepler wrote: '. . . [Brahe was] the proposer of the Rudolphine Tables, recorder of a thousand fixed star positions, observer of all planets for 38 years and continuously for 20 . . .'[19]

In the history of astronomy, however, Brahe is best known for his planetary system – a compromise between the Ptolemaic and the Copernican. First sketched out in his book on the comet of 1577, Brahe's model had a stationary Earth at the centre of the universe. The Moon was closest to the Earth, completing its orbit in 29½ days, and the Sun was farther out, orbiting the Earth in about 365 days. The Sun, however, was the centre of the circular orbits of the rest of the planets in the usual order: Mercury, Venus, Mars, Jupiter, and Saturn. The material world extended no farther than the orbit of the Moon; beyond was an ethereal realm of no mass. There were no crystalline spheres. Because of the accuracy of his observations, Brahe had to introduce a large number of epicycles and eccentrics to capture the details of his system. As a compromise the Tychonic model was a success: it avoided the main criticism of the Ptolemaists – a moving Earth – while meeting the chief objection of the Copernicans – the crystalline spheres.

Beyond his new model of the solar system, Brahe's contributions were primarily the result of his superior equipment and his dedicated programme of observations. He built larger and more stable quadrants, sextants, and octants. On the arcs of these instruments he employed a system of diagonal divisions. He invented a slit-pinnule for sighting celestial objects with an alidade. In general, these innovations enabled him to redetermine with superior accuracy every important astronomical constant then known, with the exception of the solar parallax. He also introduced atmospheric refraction (completely ignored by the Islamic astronomers) and constructed a reliable table of refractions.

Brahe conducted observations of the Sun, Moon, and planets for twenty continuous years. He made the first important step forward in lunar theory since Ptolemy by accurately and diligently observing the Moon's positions continuously round the sky. As a result, he discovered and determined the variation of the Moon's orbital inclination, determined the rate of regression of the nodes of the Moon's orbit, and explained the nineteen-year variation of the inclination of the lunar orbit. He constructed tables of the Sun's position to an accuracy of perhaps 10 or 20 seconds, much more accurate that the earlier Alfonsine and Copernican tables of 15–20 minutes. He determined the length of the tropical year to better than one second of accuracy and improved Copernicus's estimated value for the advance of the Sun's apogee with respect to the vernal equinox. He redetermined the value of the precession of the equinoxes and rejected trepidation. He constructed a catalogue of 777 fixed star positions with an unprecedented accuracy amounting to an average error of ±48 seconds. Later, he extended the catalogue to 1,000 stars with an average error of 24 seconds in declination and 25 seconds in right ascension.[20]

A fascinating article by Professor Sevim Tekeli of Ankara University compared the astronomical instruments in the Istanbul and Uraniborg observatories at the end of the sixteenth century. Taqi al-Din and Tycho Brahe were near contemporaries: Taqi al-Din, twenty years older, was born in 1526 and died in 1585 at age 59; Tycho, born in 1546, died at age 65 in 1601. Their state-of-the art observatories, the most advanced in the Eurasian world, were also contemporaneous: ground was broken for the Istanbul Observatory in early 1575 and for the Uraniborg Observatory in 1576. And the comet of 1577 was a critical event in the astronomical careers of both men.

For his comparison, Tekeli was able to utilise unusually complete and accurate contemporary descriptions. At the end of the sixteenth century, before the invention of the telescope, the instruments in the two observatories were remarkably similar. Although Tekeli's conclusion was that Tycho's observatory was superior, it does not appear, from a reading of the article, that the differences in the quality and precision of the instruments were all that great. In the case of the oldest instruments – those that could be traced back to Ptolemy and the early Greeks – there was very little variation. The armillary spheres in both places featured six rings with radii of more than 4 metres and the parallactic rulers and dioptras were essentially identical. The second group of instruments – those invented by Islamic astronomers – were also basically similar. In both Istanbul and Uraniborg the mural quadrant, first constructed by Tusi at Maragha, was set in a wall and aligned with the local meridian. Taqi al-Din's was larger – 6 metres to Tycho's 2 metres – but Tycho's had transversal markings that allowed readings to five seconds. Both observatories also featured an azimuthal semicircle (first seen in Maragha) and a wooden quadrant – Taqi al-Din's had a radius of 4.5 metres and Tycho's was larger at 5.5 metres. The third group of instruments – those invented by the two astronomers – included the sextant and the observational clock. Both clocks were extremely accurate – featuring three dials, measuring time in hours, minutes, and seconds.[21]

Tekeli's conclusion that Tycho's instruments were superior had less to do with their design and build and more to do with their use. Despite its size and staffing and its state-of-the-art equipment, the Istanbul Observatory was short-lived – finished in 1577, it was torn down at the order of Murad III in 1580. And although Taqi al-Din and his men had probably begun observations in 1573, the *zij* that the astronomer eventually compiled – *Culmination of Thoughts in the Kingdom of Rotating Spheres* – could not have gone very far in accomplishing its objective: a revision of Ulugh Beg's *Zij al-Sultani*. Tycho, on the other hand, with the lavish support of Frederick II, was able to conduct daily, systematic observations for nearly twenty years and, as a result, his discoveries were substantial. The comet of 1577 was a good example. In Istanbul it offered Taqi al-Din an opportunity to firm up Murad III's support. Its appearance heralded an Ottoman victory over the hated Safavids, confirming the value of the new institution. At Uraniborg, on the other hand, Tycho carefully observed the comet for its entire life, studying its behaviour and pondering its significance. In his 1588 treatise he argued that the comet's path through the heavens disproved the Aristotleaan theory of crystalline spheres, thereby opening the door for a physical theory of planetary motion.

Tycho's long years of observations and his careful amassing of data enabled him to propose a new planetary system and to begin what would be the most accurate set of astronomical tables in the world. The Tychonic system, with its stationary Earth and heliocentred model for the remaining four planets, was a revolutionary step forward. The first carefully-worked out challenge to the Ptolemaic model, it allowed for more accurate predictions of planetary motions and was a necessary step toward the fully-realised Copernican model of Johannes Kepler. The final product of the Uraniborg Observatory, and perhaps the real reason for Tekeli's judgment that its instruments were superior to those of the Istanbul Observatory, was the astronomical treatise – the *Rudolphine Tables* (1627). Begun by Tycho in 1601, just a few months before his death, it was finally finished by Johannes Kepler nearly a quarter of a century later. There is no doubt, however, that its great superiority over all existing treatises lay in its thorough, painstaking, and continuous collection of data amassed by Brahe and his collaborators at Uraniborg. The tables were fully Copernican, utilising Kepler's theories on the elliptical paths of the planets, and it was almost surely the contrast between the Danish and Ottoman treatises (the elegance, accuracy, and breadth of the former as compared to the latter) that prompted Tekeli to come down so strongly on the side of Uraniborg. Nevertheless, it does appear that a comparison of the observatories at Istanbul and Uraniborg – in instruments, observations, and theories – enables us to mark the moment at which leadership in astronomy and mathematics passed from the Islamic world to the European.

Johannes Kepler (1571–1630) was born near Stuttgart. His grandfather was mayor of his small hometown, but his father was disreputable – a mercenary who eventually abandoned his wife. Kepler's life was an uneasy blend of the rational and the irrational, the scientific and the superstitious. He was a practicing astrologer who defended his mother on a charge of witchcraft, but he was also a painstaking observer and theoretician who eventually worked out the laws of planetary motion.[22]

He became fascinated with astronomy at an early age – having witnessed the comet of 1577 at age six and a lunar eclipse at age nine. He attended the University of Tubingen, studying philosophy and theology, but falling under the influence of the Copernican astronomer and mathematician Michael Mastlin. He mastered both the Ptolemaic and Copernican systems and became a passionate defender of the heliocentric model. At Tubingen also, he became a skilled astrologer, casting horoscopes for his fellow students. After taking a master's degree, he began to study theology but at the death in 1594 of a mathematics professor at the Lutheran school in Graz he was asked to take over. In 1595 he published a calendar with unusually accurate weather predictions. His reputation made, he produced annual versions until 1606 and later from 1618 to 1624. As a practicing astrologer, he cast more than 800 horoscopes. In 1598 in a letter to Mastlin he identified himself a Lutheran astrologer.

In 1596, Kepler brought out his first major astronomical work, *Mysterium Cosmographicum* (*The Cosmographic Mystery*). The first published defence of the Copernican system, this treatise owed its inspiration to Kepler's fascination with astrology. For Kepler the universe was a living body – the Sun was its heart and the Earth was its liver or spleen. The planets worked within a finely balanced equilibrium. A conjunction of Saturn and the Sun caused cold weather while the conjunctions of Jupiter and Saturn, according to the theories of Abu Ma'shar, signalled important turning points in political or religious history. During a lecture on the Jupiter–Saturn conjunctions, Kepler had an important insight. The central position of the Sun, he realised, was the key to understanding planetary motion. He knew that the orbits of the planets lengthened with distance from the Sun and although the law that he promulgated to express this relationship had to be later revised, it offered substantial support for the Copernican system – the arrangement of the planets, their distance from the Sun, and the length of their orbits. He sent copies of his treatise to Galileo and Tycho. Although Galileo acknowledged privately (in a letter to Kepler) his agreement with the Copernican worldview, he feared the ridicule that would follow if he declared his allegiance openly. Tycho also admired the *Mysterium* but it was so different in style from his own empirical, observational approach that he did not at all feel threatened.

In September 1598 a Catholic commission of the Counter Revolution in Graz ordered all Lutheran teachers to leave the city. In early 1600, Kepler arrived in Prague and was invited by Tycho to work on Mars. After about a year and a half of collaboration, Tycho died, and Kepler succeeded him as imperial mathematician. His assignment was to serve as astrologer to Rudolph II and to complete Tycho's work. With complete access to the Danish astronomer's unparalleled collection of data, he set about working out the theories that would make him famous.

In 1609 he published *Astronomia nova* (*A New Astronomy*). A monumental revision and extension of the astronomy of his day, this treatise grew out of his attempts to construct a satisfactory theory of planetary motion for Mars. In it he enunciated the first two of his three laws of planetary motion. Longomontanus, Tycho's chief assistant, had tried to work out the planet's orbit without straying from the assumptions of the geocentric Tychonic system. His model employed a double epicycle and, like Copernicus, he had the planets orbiting around the centre of the Earth's orbit rather than around the true Sun. Although Longomontanus reduced the computational errors to about 2 minutes of arc, Kepler was not satisfied. Eventually, he moved the Sun to centre of his model and by postulating an ellipse instead of a circle for the planet's orbit, he achieved a much higher degree of accuracy. The most important results of this long and laborious process were his first two laws: (1) all planets move in ellipses with the Sun at one focus; (2) in their orbits the planets sweep out equal areas in equal times.

Kepler had worked on *Harmonices Mundi (Harmonies of the World)* since 1599. Finally published in 1619, it was a continuation (and in many ways a repetition) of the astrological speculations and theories of *Mysterium Cosmographicum*. He explained the astronomical and astrological aspects of the natural world in terms of harmonies – the music of the spheres. This was a venerable approach – taken up by both Pythagoras and Ptolemy, among others. He analysed first of all the harmonies among geometric shapes (regular polygons and regular solids) and then moved on to the harmonies in music, meteorology, and astrology – in astrology the harmonies between the tones of the heavenly bodies and those of the human soul. Most importantly, however, for astronomy he was able to enunciate his third law of planetary motion: the squares of the periodic times of the planets' orbits are proportional to the cubes of their mean distances from the Sun. Before this, the assumption had been that the forces between the Sun and the planets remained constant but Kepler was able to show that this power, whatever it was, decreased with distance.

Kepler published the final version of *Epitome astronomiae Copernicanae (Epitome of Copernican Astronomy)* over the period 1618–21. A textbook on the Copernican system in the form of questions and answers, it contained the first clear statement of all three of his laws. In the *Epitome* he revolutionised astronomical thought by demonstrating that the planets moved with non-uniform velocity in elliptical orbits rather than with the perpetual uniform circular motion assumed by all earlier astronomers – from Ptolemy through Copernicus, Tycho, and Galileo. The Epitome became Kepler's most influential work as it attempted to explain heavenly motions through physical laws. It was placed on the *Catholic Index Liborum Prohibitorum (Index of Prohibited Works)* in 1633 and not removed until 1822.

Although Kepler's three laws correctly described the geometry of planetary motion, he did not yet have a physical theory of causation. In his attempt to construct such a theory he brought to bear an intensely personal blend of mathematics, physics, philosophy, and mysticism. He wanted to rid his model of the multiple circles and epicycles of his forbearers (from Ptolemy to Copernicus), and he depended heavily on aesthetic considerations of simplicity, harmony, and elegance.

At first Kepler thought that planets moved because of their intrinsic powers or mental processes; a planet moved because it could judge the distance from the Sun by taking note of the Sun's apparent size. Later Kepler decided that magnetism accounted for planetary motion. The magnetic character of the Earth caused it to wobble on its axis, and the magnetic fibres in the planets, attracted or repelled by the magnetic fibres of the Sun, caused their motions. According to Kepler, the celestial machine was not a divine being but a kind of clock. It was dependent on a single driving force (magnetism) that was corporeal and physical, like the pendulum in a clock.

Kepler's last major achievement in astronomy was the *Rudolphine Tables*. Begun by Tycho Brahe soon after his move to Prague in 1599, the treatise was entrusted to Kepler after his employer's death. Although Kepler worked on these tables intermittently for many years, he never found the time to complete the final manuscript until the very end of his life. In 1612, after the death of Rudolph II, he moved to Linz, but the final manuscript was not published until 1627, three years before his own demise.

The *Rudolphine Tables* were intended to be an updated revision of the treatises current at the time. Until the end of the sixteenth century the *Alfonsine Tables* held pride of place among European astronomers and astrologers. In 1551, however, the astronomer Erasmus Reinhold published the *Prutenic [Prussian] Tables*. Based on the *De revolutionibus* of Copernicus, this treatise eventually came to replace the *Alfonsine Tables* through most of Europe and served to publicise to a much wider audience the revolutionary heliocentric model. The *Rudolphine Tables*, however, represented a quantum leap in accuracy and ease of use. Based primarily on the voluminous and meticulous compilation of data amassed by Brahe over the course of his long career, the Rudolphine treatise did not adopt the geocentric Tychonic model. Rather, Kepler replaced Brahe's model with his own – heliocentric, the planets revolving around the Sun in ellipses. The result was the most accurate set of astronomical tables yet. In addition to the increased accuracy of the data, the use of the newly discovered system of logarithms made the calculations of local astrologers much less laborious. The increased accuracy of the tables was significant – in the case of Mars errors of five degrees were reduced to errors of a few minutes only. Kepler was able, for example, for the first time to predict the transits of Mars and Venus across the Sun's disk. Of Kepler's achievement, one historian of science wrote:

> . . . at one stroke he had blown away all the ancient cobwebs of about 240 eccentrics, epicycles, and equants that described motions on motions about mathematical points having no physical existence, but serving merely as a mental convenience for calculation.[23]

In the history of astronomy, Galileo Galilei (1564–1642) stands between Kepler and Newton. A physicist, mathematician, and astronomer, he is best known for his technical work on the early prototypes of the telescope and for the observations that helped to solidify the evidence in favour of Copernicus. Born in Pisa, Galileo was sent to the University of Pisa to study medicine but became fascinated by mathematics. After graduation and a job in Florence, he was appointed to the chair of mathematics in Pisa. Four years later he moved on to the chair of mathematics at the University of Padua.

Galileo was the first astronomer of note to make extensive use of the telescope. Although he did not invent the instrument, he made significant improvements in the earliest models. In 1609 he built telescopes with magnifications of two to three times but within a year he had improved the magnification to twenty to thirty. He observed the mountains on the Moon, the four moons of Jupiter, and the starry constitution of the Milky Way. He reported his discoveries in his 1610 book, *Sidereus nuncius* (*Sidereal Messenger*). In 1613 he published a book on sunspots. In subsequent years he detected the phases of Venus and what were later to be called the Rings of Saturn.

Galileo was immediately aware of the implications of his discoveries. They seriously undermined the tenets of Aristotelian cosmology. The appearance of the supernova of 1604, the number of stars in the Milky Ways, the moons of Jupiter, and the rings of Saturn – all appeared to contradict the perfect and incorruptible universe of the great philosopher. In 1611 he presented his telescopic findings to the Jesuits. He defended heliocentrism, claiming that it was not contrary to the scriptures. Like Augustine, he argued that passages from books of poetry or songs should not be taken literally. The Biblical authors were writing from a terrestrial perspective and from that vantage point the Sun does rise and set. His argument, however, was unsuccessful, and in 1616 he was ordered by a Papal commission in Rome to abandon the idea that the Earth moved and to cease defending it. A week later Copernicus's *De revolutionibus* was put on the Prohibited List. For the next several years he followed the commission's bidding, but in 1623 his friend and admirer, Cardinal Maffeo Barberini, was elected Pope Urban VIII. Barberini had opposed the condemnation of Galileo in 1616 and so the astronomer went back to work. In 1632 he published *Dialogue Concerning the Two Chief World Systems*. Asked by Urban VIII to give arguments for and against heliocentrism and not to favour the latter, Galileo gave the name Simplicio (vulgarly, Simpleton) to the geocentrist Ptolemian and made him appear ridiculous. In 1633 he was called to Rome by the Inquisition. Although the examination was held under threat of torture, there was in all likelihood, no intention to resort to actual violence. Throughout the process Galileo maintained that he had given up his Copernican views after 1616. Judged guilty of heresy (for holding that the Sun lies motionless at the centre of the universe), he was sentenced to life imprisonment. His *Dialogue* was banned and publication of his writing was forbidden. His sentence was later commuted to house arrest, and he lived another eleven years, falling blind four years before his death. His staunch defence of Copernicus, however, did not include the Keplerian ellipses, and he remained true to the complicated Tychonic system of cycles and epicycles to the end of his life.[24]

It is entirely appropriate that this discussion of astronomy and astrology in the Islamic world end with Isaac Newton (1642–1727). For it is only with the great English mathematician and physicist that we come full circle to an understanding of the difficulties faced by the Egyptians, Greeks, and Muslims in tracking and explaining the movements of the heavenly bodies. It is a long way from the rotating crystalline spheres of Aristotle to a solar system of massive planets orbiting the Sun in ellipses determined by the Englishman's law of universal gravitation.

It was the unprecedented accuracy of the *Rudolphine Tables* that caused even the most hidebound European astronomers to take seriously the Copernican system. But, as we have seen, Kepler's laws were purely descriptive. No material, physical theory was advanced to account for the movements of the planets. After Kepler, the first attempt at a physical theory – one in which matter acted on matter – was provided by the French philosopher and mathematician René Descartes (1596–1650). His theory of celestial vortices (or whirlpools of subtle matter) argued that the material world consisted of three kinds of matter: luminous, transparent, and opaque. The luminous, the finest and fastest, consisted of particles moving at great speed and was the material of the Sun and stars. The opaque was the coarsest, heaviest, and slowest and made up the Earth and planets. Transparent matter filled the spaces between the celestial bodies. According to Descartes, it was the magnetism created by the spinning vortices that caused the motions of the heavenly bodies.

Isaac Newton was born in 1642 and entered Trinity College, Cambridge in 1661. Interested in mathematics and science, he studied the laws of Kepler and the theories of Descartes. Impressed by his precocious genius, the university appointed him Lucasian Professor of Mathematics in 1669. He remained at Cambridge until 1696 when he was made Warden of the Mint. Like a medieval Islamic polymath, he wrote on a variety of subjects: mathematics, physics, optics, and theoretical mechanics in addition to religion and alchemy. In 1664 he put forward his law of universal gravitation – any two bodies in the universe attract one another with a force that is directly proportional to the product of their masses and inversely proportional to the square of the distance between them. At the end of the 1670s and early 1680s, with his discoveries in mathematics and physics behind him, Newton turned to Kepler's third law, the law of areas. Challenged to explain how a law of central force could turn the straight line motion of a planet into a Keplerian ellipse, he showed that the law of areas implied that the force was directed to a single centre and was the inverse square of the distance. He also showed that the orbits of comets were parabolic (open ellipses).

In Book Three (*The System of the World*) of his *Principia mathematica philosophiae naturalis* (*Mathematical Principles of Natural Philosophy*), first published in 1687, Newton gave the first complete explanation of material movement in all parts of the universe as the action of a single set of laws. The motions of the planets and their satellites, the motions of the comets, the Earth, and the ocean tides – all were explained by the law of universal gravitation. All matter attracts other matter. The Sun attracts the planets and the planets the Sun. The force is independent of the kind of matter: only the quantity and separation are important. Although the gravitational force of the Sun on the planets is greater than the force between the planets themselves, the latter cannot be ignored, especially when the planets pass close to one another. The disturbing action of the Earth on the Moon must be also taken into account. Newton showed that the gravitational force of the Sun, slightly greater on the near side of the Earth, caused its flattened shape. This discrepancy produced a turning effect (or couple) on the axis of the Earth which caused the precession of the equinoxes.[25]

As in the Islamic world, so too in the early European, astrology was a handmaiden to astronomy, claiming the attention of even the most talented astronomers and mathematicians. In astrology also, the early stimuli originated in Islamic Spain. In 1142 John of Seville translated a collection of Arabic texts (those of Abu Ma'shar especially) into Latin with the title *Epitome of the Whole of Astrology*.[26] Other scholars followed his lead and by the first decade of the thirteenth century the complete works of Aristotle (first encountered in the writings of the Islamic astronomers) were available in Western Europe in a language that every scholar could read. And the underlying assumption of Aristotle's cosmology – that the processes of change and growth on Earth depended on the activities of the heavenly bodies – was everywhere accepted, giving to astrology an unshakeable justification. From this time forward philosophers, theologians, and poets – from Albert Magnus and Thomas Aquinas to Dante, Chaucer, and Shakespeare – never questioned the belief that the planets had a major influence on human affairs. And since the medieval period in Europe was a time of crisis (the Black Death, the long wars between France and England, the fears of the Antichrist, and the Protestant Reformation), men of influence began to look to astrology as a way of explaining the natural, social, and political upheavals of the time.[27]

Albertus Magnus (1193–1280), one of the greatest scholastic philosophers and the teacher of Thomas Aquinas, shared the common opinion of the planets' influence and, though he defended free will, he nevertheless asserted that a properly trained astrologer could make accurate predictions about the life of an infant, within the limits of what God allowed. Aquinas, like Albertus, denied that the stars were living beings but claimed that no intelligent man could doubt that the natural motions of the inferior bodies were caused by the movements of the planets and stars. Although Roger Bacon (1214–94) mounted a violent attack on magic and magicians, he wholeheartedly agreed with the astrology of Albertus and Aquinas. Bacon maintained that the heavenly bodies could incline men to good or evil, and he wrote about the planets and their influence on the course of Christian history. Dante Alighieri (1265–1321) had read the works of those who defended astrology, and although he condemned some astrologers in the *Inferno*, in the *Paradiso* he celebrated astrology as the interpreter of the will of God. Geoffrey Chaucer (c. 1345–1400) was the first English writer whose works were saturated from beginning to end with astrology. He wrote a treatise on the astrolabe, making use of Masha'allah's commentary on Ptolemy, and in his long poem *Troilus and Criseyde* he included carefully calculated astronomical allusions. In the *Canterbury Tales* (especially in *The Man of Law's Tale* and the *Wife of Bath's Tale*), horoscopes played an important role. The plays of William Shakespeare were also shot through with references to astrology: horoscopes, stars, eclipses, comets, and conjunctions.[28]

In Western Europe the rise of printing (invented by Johann Gutenberg in 1454) gave birth to an ever-widening interest in astrology, expanding its reach from the educated elite to the general public. In 1467 Regiomontanus composed his *Table of Directions* – a treatise on how to divide the ecliptic into the twelve houses of the zodiac – and in 1474 he printed his *Kalendarium and Ephemerides* – a calendar/almanac that became immensely popular. His work included calendars with anniversaries, festivals, weather forecasts, medical information, and astronomical data. It also included astrological notes on blazing stars, comets, diseases, plagues, freaks of nature, prophecies, and predictions.[29] These almanacs became widely distributed first in Germany and the Low Countries but they soon spread throughout Europe. For the barely literate, they were often the only printed material in the house. The first printed English almanacs came out in 1500, and in 1603 James I granted a monopoly on almanac, prayer book, and Psalter printing to a joint stock company. By the late seventeenth century in England 400,000 almanacs were sold annually.[30]

With the wide availability of almanacs came a renewed interest in horoscopes. Books containing the nativities of famous men became increasingly popular – a 1543 work contained sixty-eight biographies and was later expanded to include 100. Casebooks of famous individuals as character types were quickly printed and immediately snapped up. A subgenre that appealed to both the pious and the superstitious centred on the horoscopes of Jesus Christ. For the ordinary medieval scholar it must have seemed that Matthew's story of the wise men and the star sanctioned an astrological interpretation of the Christmas story. And over the centuries a number of such explanations surfaced. Pietro d'Abano (1257–1316), an Italian astrologer, suggested that a conjunction of Jupiter and Saturn (following Abu Ma'shar) had foretold the rise of Christianity. And Cardinal Peter d'Ailly (1351–1420), a French theologian and astrologer, asserted that the stars had exerted an important impact on the lives of Christ and his mother, shaping the development of their natural virtues. In 1465 Jacob von Spier asked Regiomontanus about the relationship between astrology and Christianity: is it possible to determine the year of Christ's birth from the horoscope of the preceding great conjunction?[31]

By the end of the sixteenth century, however, the tide had begun to turn and genuine interest in astrology began gradually to diminish. The principal factors behind this development were the new discoveries in astronomy and physics. The universal realisation that the Sun rather than the Earth was at the centre of the universe seemed to devalue the underlying principle of astrology. And the acceptance of the great distance between planets (and the even greater distance between the stars and planets) made it extremely difficult to imagine that any significant influence could be actually exerted. Finally, there was the growing conviction that any scientific idea should be capable of technical explanation. The tell-tale sign of the astrology's fall from grace came in the last years of the century, heralded by a series of Popes who were actively opposed to the practice. In 1586, for example, Sixtus V enacted a bull against judicial astrology. The casting of horoscopes was also prohibited: only God could know the future.[32]

Notes

1. Stephen C. McCluskey, *Astronomies and Cultures in Early Medieval Europe* (New York: Cambridge University Press, 1998), 165–8.
2. Blake, *Time*, 179–80.
3. Blake, *Time*, 174; McCluskey, *Astronomies and Cultures*, 163–5.
4. North, *Cosmos*, 142–7, 251–4; McCluskery, *Astronomies and Cultures*, 115–17.
5. North, *Cosmos*, 222.
6. Ibid. 222–3.
7. Ibid. 224–6, McCluskey, *Astronomies and Cultures*, 168–87.
8. North, *Cosmos*, 227–9.
9. Ibid. 228–30.
10. Ibid. 250–5.
11. Ibid. 255–8.
12. Ibid. 258–64.
13. Ibid. 264–6.
14. Ibid. 270–7.
15. Ibid. 302–20; Jacobsen, *Planetary Systems*, 103–49.
16. George Saliba, *Islamic Science and the Making of the European Renaissance* (London & Cambridge, MA: MIT Press, 2007), 209.
17. Ibid. 193–232.
18. Jacobsen, *Planetary Systems*, ch. 6.
19. Ibid. 168.
20. North, *Cosmos*, 321–38; Jacobson, *Planetary Systems*, ch. 6.
21. Sevim Tekeli, 'Astronomical Instruments of Tycho Brahe and Taqi al-Din', in Muammer Dizer (ed.), *Proceedings of the International Symposium on the Observatories in Islam (19–23 September 1977)* (Istanbul: Milli Egitim Basimevi, 1980), 33–43.
22. Jacobsen, *Planetary Systems*, 172–254; North, *Cosmos*, 332–60.
23. Jacobsen, *Planetary Systems*, 176.
24. North, *Cosmos*, 360–80.
25. Ibid. 399–417.
26. Ibid. 286–92.
27. Ibid.
28. Derek & Julia Park, *A History of Astrology* (London: André Deutsch, 1983), 101–2.
29. North, *Cosmos*, 292–301.
30. Ibid. 292–4.
31. John David North, *Horoscopes and History* (Warburg: Warburg Institute, 1986), 158–66.
32. Park, *A History of Astrology*, 244.

CHAPTER TEN

Conclusion

In the popular story of astronomy the 1,400 years between the geocentric system of Claudius Ptolemy and the heliocentric model of Nicholas Copernicus was a period of no great consequence. Muslim astronomers translated the *Almagest*, the *Handy Tables*, and the *Tetrabibios* from Greek into Arabic in eighth century Baghdad and, after a few editorial comments and suggestions, saw the Arabic versions translated into Latin in eleventh century Spain and passed on to the fledgling scholars of late medieval Europe. In this account Muslim astronomers and mathematicians functioned primarily as conduits, transmitting a relatively untouched version of the famous Alexandrian, adding little to a model that was questioned, criticised, and finally overturned by the astronomer/ mathematicians of Renaissance Europe – Nicholas Copernicus, Tycho Brahe, Johannes Kepler, and finally Isaac Newton.

Although this cartoon version has been challenged by serious scholars of Islamic history and science, a non-technical account of the Muslim contributions to astronomy and astrology has not been available. To fill that void and to highlight the crucial role that Islamic scientists played in the transition from Ptolemy to Copernicus and Newton has been the goal of this essay. Astronomy was in many ways the first science and the effort to untangle the mysteries of the heavens led to a number of important discoveries in mathematics, geometry, trigonometry, and physics. Because of the complex gravitational forces in the solar system – the massive Sun acting on the five smaller but distant planets – the task was incredibly difficult. And it was the effort to decipher the riddles of the heavens that spurred the earliest efforts to understand the world in a systematic, proto-scientific manner – to count, model, and predict.

In the wider context of Eurasian science the Islamic achievement was central, connecting the first century Greeks and their predecessors (the Egyptians and the Babylonians) to the Renaissance world of sixteenth-century Europe. And from this perspective, the breakthroughs in European astronomy, mathematics, and physics would not have been possible without the work of the Muslim astronomers and mathematicians who went before – men like Abu Ja'far Muhammad al-Khwarizmi, Nasir al-Din Tusi, and Jamshid al-Kashi.

Of the early Muslim astronomer/astrologers the most important were al-Khwarizmi and Abu Ma'shar. Al-Khwarizmi introduced the Hindu (misnamed 'Arabic') numeral system (numbers 1–9, 0, and the place value system) and algebra (the branch of mathematics that studies symbols and their manipulation) to both the Islamic and the European worlds. And both of these were absolutely essential for the breakthroughs in the higher reaches of astronomy, mathematics, and physics. Al-Khwarizmi's work was popularised in Andalusia by al-Majriti and then passed on to Europe in the Latin translations of Majriti's version of Khwarizmi's *Zij al-Sindhind*. And Abu Ma'shar added to the basic ideas of Ptolemy's *Tetrabiblos* the concept of historical astrology – the Iranian theories about the political and religious significance of the various Jupiter–Saturn conjunctions.

An accidental but important feature of astronomy in the Islamic world was its connection to Spain and through Spain to medieval Europe. Although Andalusia was an early outpost of Islamic expansion (the Spanish Umayyads establishing themselves in Cordoba in the early eighth century), this Western-most base of Islamic rule was soon cut off from the eastern Islamic world by the Reconquista of the eleventh–thirteenth centuries. And this was crucial because it meant that the great accomplishments of Islamic astronomy – the work of Nasir al-Din Tusi at Maragha and Ulugh Beg at Samarqand – were virtually unknown in Andalusia and were thus unavailable to the fledgling scientists of medieval and early Renaissance Europe.

The astronomical observatory was an Islamic creation. The establishment of a scientific institution – with guaranteed funding, a library, a collection of instruments, a staff of astronomers and mathematicians, a group of students, and a mission – all of this was an Islamic first. And it was in these observatories that the cutting-edge work of the Islamic astronomers and mathematicians took place. The first observatories were modest – a few instruments, several astronomers, and a short life. But the Maragha Observatory, financed by the Mongol ruler Hulagu and directed by Nasir al-Din Tusi, was a splendid affair – the largest, most sophisticated, and productive scientific institution in the entire Eurasian world. And it was famously productive. Tusi, its director and guiding genius, was one of the most remarkable of Islamic polymaths. Astronomer, mathematician, theologian, and philosopher, he was responsible for the 'Tusi Couple', the only new mathematical theory of medieval astronomy. A major influence on the planetary systems of both Regiomontanus and Copernicus, it offered a solution to one of the major problems of Ptolemy's *Almagest*, allowing astronomers to retain the accuracy of Ptolemy's mathematical predictions while avoiding the physical contradictions of his planetary model. In computational mathematics, Tusi combined the Greek and Indian numeral systems and in pure mathematics he established trigonometry as a separate discipline, providing a description of spherical trigonometry that was independent of astronomy. Mu'ayyad al-Din 'Urdi, another Maragha astronomer, supervised the construction of the observatory and devised a mathematical theorem that transformed the eccentric motions of the planets in Ptolemy's model into epicycles, thereby ridding the *Almagest* of another contradiction. Copernicus employed the "Urdi Lemma' in constructing his model of the upper planets. A comparison of the two astronomical treatises – the *Zij-i Ilkhani* and the *Alfonsine Tables* – revealed the overwhelming superiority of Tusi's creation and the great gap between the work of the Islamic astronomers and mathematicians in the East and that of the Spanish and early European astronomers in the West. And this gap would not be closed for the next 250–300 years.

The other great observatory in the Islamic world was in Samarqand. Like Maragha, the institution founded and funded by the Timurid prince, Ulugh Beg, was the finest scientific institution of its time. Coupled with a madrasa dedicated to the rational sciences, boasting a talented faculty, a large student body, and an extensive library, the observatory contained the largest and most precise collection of observational instruments in the Eurasian world. The leading astronomer and mathematician was the brilliant Jamshid al-Kashi and his contributions were decisive. In his *Treatise on the Circumference* he calculated the value of π to sixteen decimal places – Ptolemy had achieved five places and a Chinese scholar six – but it was not until the late seventeenth century that al-Kashi's estimate was surpassed. In *Key to Arithmetic* al-Kashi composed a methodical introduction to decimal fractions, showing how arithmetical operations could be carried out on fractions as well as on integers. He also explained the analogies between the sexagesimal and decimal systems of fractions. Finally, in his *Treatise on the Chord and Sine* he calculated the value of sine 1 degree to 10 sexagesimal places (the previous value had been to only four places). The *Zij al-Sultani*, when finally finished, was the most accurate of its time – a new star catalogue and tables of greater precision. This was due both to al-Kashi's theoretical innovations and to the superiority of the observational instruments – much larger and thus much more precise.

Although the story of Islamic astronomy and astrology does not end with Samarqand, Ulugh Beg's observatory marked its climax. The Istanbul Observatory of Taqi al-Din, constructed a century and a half later, boasted a sophisticated collection of instruments and a talented director. It was well enough equipped to push the boundaries of the science and to maintain the superiority of eastern astronomy. However, unlike Nasir al-Din Tusi at Maragha or Jamshid al-Kashi at Samarqand, Taqi al-Din did not have the full support of his patron. After barely three years of operation, the instruments and the buildings were razed, victims of a dynastic power struggle. The abrupt collapse of the Istanbul Observatory, so soon after its founding, contrasted with the longevity and productivity of the contemporaneous European institution at Uraniborg. With Tycho Brahe's observatory, founded the same year as the Ottoman institution, leadership in Eurasian astronomy and mathematics passed from the Islamic East to the Christian West. In its instruments, library, staff, and support facilities, Uraniborg was far superior to the Istanbul Observatory, and

Tycho's observations – detailed, precise, and protracted – laid the groundwork for the final acceptance of the Copernican revolution.

The last observatory (observatories) in the Muslim world served only to underline the transition that had occurred. By the mid-eighteenth century the heliocentric system of Copernicus, buttressed by the theories of Kepler and Newton, and supported by the telescopic discoveries of Galileo and others, had swept the field. The astronomy of Maragha and Samarqand and their treatises – the *Zij-i Ilkhani* and the *Zij al-Sultani* – had been surpassed. But Jai Singh, although he had access to European astronomers and astronomical treatises, was not able to change direction. His observatories were filled with instruments of masonry, his telescopes were rudimentary, and his treatises knew nothing of Copernicus, Kepler, or Newton. But this rather feeble end to the story of Islamic astronomy and astrology should not blind us to the great achievements of its earlier period. From al-Khwarizmi in eighth-century Baghdad until Taqi al-Din in sixteenth-century Istanbul, the exciting discoveries in astronomy, mathematics, and astrology were found in the Islamic world – not in the European.

Glossary: astronomical instruments

Just as Muslim astronomers and mathematicians led the way in the development of planetary models and mathematical and trigonometric theories, they were also primarily responsible for the improvements in astronomical instrumentation in the 1,400 years between Ptolemy, on the one hand, and Tycho Brahe and Galileo Galilei, on the other. There were two main categories of Islamic instruments: the observational and the non-observational. For observing the heavens Muslim astronomers employed, for the most part, three instruments: the armillary sphere, the mural quadrant, and the parallactic ruler. The main purpose of the non-observational instruments, on the other hand, was to solve problems in spherical astronomy. These involved the mathematics of celestial configurations and consisted primarily of problems related to timekeeping. The rising and setting of the Sun and stars over the local horizon were the main indicators. In this category were four instruments: the celestial sphere, the astrolabe, mathematical grids, and sundials.

Armillary sphere

The armillary sphere was a spherical framework of rings centred on the Earth and intended to be a model of celestial objects. The rings represented astronomically significant circles on the celestial sphere – the horizon, the meridian, the celestial equator, and the ecliptic. The armillary sphere was used to locate the positions of the stars and planets at a given time and place. The Greek astronomer, Hipparchus, credited Eratosthenes (276–194 BCE) with the invention of the device. In Latin *armilla* meant 'circle', and in its simplest form the instrument consisted of a ring fixed in the plane of the equator – an equinoctial armilla. Adding another ring in the plane of the meridian transformed it into a solstitial armilla. It became an armillary sphere with the addition of more rings – representing the horizon, the celestial equator, and the ecliptic. Eratosthenes probably used the solstitial armilla to measure the obliquity of the ecliptic. Ptolemy further refined the instrument he had inherited, adding two small sighting tubes. In the early eighth century the Islamic astronomers of Baghdad fabricated larger, more complicated, and more precise instruments. Introduced to Western Europe by way of Andalusia in the early eleventh century, the armillary sphere became a staple of early Renaissance astronomy. Tycho Brahe constructed an elaborate version of the early Islamic model. These instruments were among the first complex mechanical devices, and Renaissance scientists and scholars often had their portraits painted holding an armillary sphere in one hand – symbolising their mastery of the science and technology of their day.[1]

Astrolabe

The astrolabe (from the Greek 'star taker') was invented in ancient Greece and perfected by Hipparchus in the second century BCE. Although used by Ptolemy, the instrument was elaborated and refined in the Islamic world, before finally reaching Western Europe by way of Andalusia in the eleventh and twelfth centuries. An incredibly versatile instrument, it had, according to al-Sufi, more than one thousand functions – it was rightly called 'a universe in one's palm'.

The astrolabe was an analogue computer that represented in two dimensions – rather than three – the positions of the Sun, Moon, planets, and fixed stars with respect to the local horizon. To accomplish this it employed stereographic projection in two ways: first, the celestial circles were projected on the astrolabe plane as circles; and second, the angles between the intersecting celestial circles remained unchanged when projected on the plane. The most common type, the northern astrolabe, had the North Pole as its centre and the Tropic of Cancer as its outermost periphery. This version could only show stars that lay north of the Tropic of Cancer.

The astrolable consisted of three parts. The mater (mother) was a disk which was deep enough to hold one or more flat plates called tympans (climates). Each tympan was made for a specific latitude and was engraved with a stereographic projection of circles denoting azimuth and altitude and representing the portion of the celestial sphere above the local horizon. The rim of the mater was graduated into hours of time or degrees of arc or both. The third part of the astrolabe was the rete. This was a framework bearing a projection of the ecliptic plane and pointers for stars. The rete was a star chart and could be rotated. One complete rotation over the projection on the tympan corresponded to the passage of a single day. On the back of the mater several scales were engraved: curves for time conversion, a calendar for converting the day of the month to the Sun's position on the ecliptic, trigonometric scales, and a graduation of 360 degrees around the back edge. An additional piece, the alidade (or sighting device), was often attached to the back face. When the astrolabe was held vertically, the alidade could be rotated until the celestial object and its altitude in degrees could be read from the graduated edge of the mater.

The main function of the astrolabe was timekeeping. With its help time could be determined both during the day and at night, in equal hours and/ or in seasonal hours. Equal hours were those obtained by dividing the nychthemeron (night and day) into twenty-four equal parts, that is, hours of 60 minutes each. Unequal or seasonal hours were obtained by dividing the periods of daylight and of darkness into twelve equal parts each. Such hours were longer in summer and shorter in winter. In the Islamic world seasonal hours were used for all practical purposes while equal hours were used for astronomical calculations.

The astrolabe could also be set to show the positions of the heavenly bodies at different times for different latitudes and could solve problems concerning the positions of the Sun, Moon, stars, and planets at a given time. As such, it was an indispensable tool for the astrologer who, in casting horoscopes, had to calculate the four points where the ecliptic intersected the horizon and the meridian. To compute these four mathematically was a tedious process. But they could be read directly from the astrolabe when the rete was set for the proper moment. When the four key points were known, the twelve astrological houses could be easily determined and the various planets assigned to them.[2]

Celestial sphere

The celestial sphere was a large sphere with the Earth at its centre. It displayed the sky as if it were projected on the underside of a large dome. It was a practical tool that allowed the astronomer to plot celestial objects whose distances were unknown. The sphere could be rotated around its axis so that the risings and settings of celestial objects over the horizon could be simulated.

Dioptra (diopter)

The dioptra was a classical astronomical surveying instrument dating from the third century BCE. It was a sighting tube or a rod with a sight at both ends. It was used by the Greek astronomers to measure the positions of stars. By the time of Ptolemy it was mostly obsolete, having been replaced by the armillary sphere. It was sometimes employed by Islamic astronomer/astrologers to measure the diameters of stars or the distance between them.

Equatorium

The equatorium was an astronomical calculating instrument that could be used to find the positions of the Sun, Moon, and planets without computation. It used a geometrical model to represent the position of a given celestial body. It had a series of rotating disks, each representing the path through the sky of a celestial body – the Sun, Moon, and the five planets. The disks were operated manually. The earliest record of a solar equatorium (one to find the position of the Sun) was found in the work of a fifth century BCE Greek astronomer. Two works were written on the instrument: a work by al-Zarqali in the eleventh century and a work composed in the thirteenth century by the English astronomer, Richard of Wallingdon.

Horoscope

A horoscope (from Greek, 'observer of the hour') was an astrological chart that represented the positions of the Sun, Moon, planets, and astrological aspects at a particular time – usually the hour of the subject's birth. It was a stylised map of the heavens over a specific location at a particular moment in time. The perspective was usually geocentric and the positions of the heavenly bodies were placed in the chart along with those of purely calculated factors – the lunar nodes, the ascendant, zodiac signs, and fixed stars. Angular relationships between the planets themselves and other points, called aspects, were often determined. The interpretation of the horoscope differed according to the tradition, skill, and experience of the astrologer. To create a horoscope, the astrologer had to first determine the exact time and place of the subject's birth. Then, using a set of tables (an ephemeris or *zij*), he ascertained the precise location of the Sun, Moon, and planets at the required time. The chart itself was divided into twelve houses, on which the twelve signs of the zodiac were superimposed. The heavenly bodies were positioned at the proper places and other relationships were often calculated.[3]

Mural quadrant

A mural quadrant was an angle-measuring device mounted on or built into a wall. It was placed precisely on the local meridian and measured the angles or altitudes of heavenly bodies from 0 to 90 degrees. The arc of the quadrant was marked with divisions which represented degrees and fractions of degrees. The larger the quadrant the finer and more precise the measurements. A mural sextant was smaller, measuring angles of 60 degrees or less. The mural quadrant was sometimes called a meridian arc.

Parallactic ruler

The parallactic ruler or *triquetrum* (*three-cornered*) served the same function as the mural quadrant – measuring the altitude of celestial objects – but it was smaller and more portable. It consisted of three wooden poles: a vertical post with a graduated scale and two pivoted arms. The star or planet was sighted along one arm and the angle was read off from the other. Ptolemy described the instrument in the *Almagest* and it was one of the most popular astronomical instruments until the invention of the telescope. Copernicus described it in his *De revolutionibus*, and it was employed by Tycho Brahe at his observatory at Uraniborg.[4]

Sundial

The sundial was a device that told the time of day by the position of the Sun. In the horizontal sundial, the most common design, the Sun cast a shadow from a gnomon (a thin rod or cylinder) onto a surface marked with lines indicating the hours of the day. As the Sun moved across the sky the shadow edge aligned with the different hour lines and the time could be read off.[5]

Water clock

The water clock, or *clepsydras* (water thief), was an early timekeeping device that did not depend on the observation of celestial bodies. The earliest examples were stone vessels with sloping sides that allowed water to drip at a constant rate from a small hole near the bottom. Other water clocks were cylindrical or bowl-shaped containers designed to slowly fill with water coming in at a constant rate. Markings on the inside surfaces measured the passage of 'hours' as the water level reached them. These clocks were normally used to determine the hours at night but they were sometimes employed during the daytime as well. More elaborate water clocks were developed by Islamic astronomers. They regulated the water pressure to achieve a more constant flow, and they added fancier time-displays: bells rang, doors or windows opened to show little people, and pointers or dials revealed astronomical and astrological information.

Notes

1. For a brief overview see Henry Smith Williams, *A History of Science* (Whitefish, MT: Kessinger Publishing, 2004; Emilie Savage-Smith, *Islamicate Celestial Globes: Their History, Construction, and Use* (Washington, DC: Smithsonian Institution Press, 1985.
2. David A. King. *In Synchrony with the Heavens: Studies in Astronomical Timekeeping and Instrumentation in Medieval Islamic Civilization*, vol. 2: *Instruments of Mass Calculaton* (London & Leiden: Brill, 2005); D. A. King, 'The Origin of the Astrolabe according to the Medieval Islamic Sources', *Journal for the History of Arabic Science* 5 (1981): 43–83; Sreeramula Rajeswara Sarma, *Astronomical Instruments in the Rampur Raza Library* (Rampur: Rampur Raza Library, 2003), 3–14; James E. Morrison, *The Astrolabe* (Janus, 2007).
3. For a general introduction see John David North, *Horoscopes and History* (London: Warburg Institute, 1986).
4. King, *Instruments of Mass Calculaton*, 12.
5. Ibid. 13.

Select bibliography

'Abu Ma'shar', s.v. *Dictionary of Scientific Biography*.

'al-Battani, Abu 'Abd Allah Muhammad ibn Jabi ibn Sinan al Raqqi al Harrani al Sabi', s.v. *Dictionary of Scientific Biography*.

'al-Biruni, Abu Rayhan Muhammad ibn al-Biruni', s.v. *Dictionary of Scientific Biography*.

'al-Biruni, Abu Rayhan Muhammad ibn al-Biruni', s.v. *Encyclopaedia Iranica*.

'al-Khayyam, 'Umar ibn Ibrahim al-Nishapuri', s.v. *Dictionary of Scientific Biography*.

'al-Khwarizmi, Abu Jafar Muhammad ibn Musa' s.v. *Dictionary of Scientific Biography*.

'al-Farghani, Ahmad ibn Muhammad ibn Kathir', s.v. *Dictionary of Scientific Biography*.

'al-Kashi (or al-Kashani), Ghiyath Al Din Jamshid Masud', s.v. *Dictionary of Scientific Biography*.

'al-Majriti, Abu al-Qasim Maslama ibn Ahmad al-Faradi', s.v. *Dictionary of Scientific Biography*.

'al-Tusi, Muhammad ibn Muhammad ibn al-Hasan, s.v. *Dictionary of Scientific Biography*.

'al-Zarqali, Abu Ishaq Ibrahim', s.v. *Dictionary of Scientific Biography*.

Ansari, S. M. Razallah. 'On the Transmission of Islamic Astronomy to Medieval India', in *The Tradition of Astronomy in India, Jyotihsastra*. Ed. B. V. Subbarayappa. New Delhi: Center for Studies in Civilisations, 2008. Pp. 345–73.

'Astrology and Astronomy in Iran', s.v. *Encyclopaedia Iranica*.

Beck, Roger. *A Brief History of Ancient Astrology*. London: Blackwell Publishing, 2007.

Blake, Stephen P. *Time in Early Modern Islam: Calendar, Ceremony, and Chronology in the Safavid, Mughal, and Ottoman Empires*. Cambridge: Cambridge University Press, 2013.

Chabas, Jos and Bernard R. Goldstein. *The Alfonsine Tables of Toledo*. Boston & London: Kluwer Academic Publishers, 2003.

Chelkowski, P. (ed.). *The Scholar and the Saint: Studies in Commemoration of Abul-Raihan al-Biruni and Jalal al-Din al-Rumi*. New York: Hagop Kevorkian Center for Near Eastern Studies, 1975.

Fitzgerald, Fitzgerald. *Rubaiyat of Omar Khayyam*. Ed. Christopher Decker. Charlottesville: University of Virginia Press, 1997.

Ghori, S. A. 'Appendix III: Development of Zijes in India', in *The Tradition of Astronomy in India, Jyotihsastra*. Ed. B. V. Subbarayappa. New Delhi: Center for Studies in Civilisations, 2008. Pp. 387–408.

Gingerich, O. *The Eye of Heaven: Ptolemy, Copernicus, Kepler*. New York: American Institute of Physics, 1993.

——— 'A Tusi Couple from Schoener's 'De Revolutionibus'', *Journal of the History of Astronomy* 15:2 (1984): 128–33.

——— 'Ibn Shatir', in *The Biographical Encyclopedia of Astronomers*. Ed. Thomas Hockey et al. New York: Springer Reference, 2007.

——— 'Ihsan Fazlioglu, Qushji, Abu al Qasim 'Ala al Din Muhammad Qushci-zade', in *The Biographical Encyclopedia of Astronomers*. Ed. Thomas Hockey et al. New York: Springer Reference, 2007.

Ihsanoglu, Ekmeleddin (ed.). *History of the Ottoman State, Society, and Civilisation*, 2 vols Istanbul: Research Center for Islamic, Art, and Culture, 2001–2.

Ileri, Ilay, "Ali al-Qushji and His Contributions to Mathematics and Astronomy', *Journal of the Center for Ottoman Studies* 20 (2006): 175–83.

Jacobsen, Theodor. *Planetary Systems from the Ancient Greeks to Kepler*. Seattle: University of Washington Press, 1999.

Kasir, D. S. (tr.). *The Algebra of Omar Khayyam*. Beirut, 1972.

Kennedy, E. S. and David Pingree. *The Astrological History of Masha'allah*. Cambridge, MA: Harvard University Press, 1971.

Kennedy, E. S., 'An Astrological History based on the Career of Genghis Khan', in *Astronomy and Astrology in the Medieval Islamic World*. Ed. E. S. Kennedy. London: Ashgate Variorum, 1998. Pp. 218–32.

——— *A Survey of Islamic Astronomical Tables*. New York: American Philosophical Society, 1956.

King, David A. *In Synchrony With the Heavens: Studies in Astronomical Timekeeping and Instrumentation in Medieval Islamic Civilization. Volume 1: The Call of the Muezzin*. London & Leiden: Brill, 2004.

——— *In Synchrony With the Heavens: Studies in Astronomical Timekeeping and Instrumentation in Medieval Islamic Civilization. Volume 2: Instruments of Mass Calculaton*. London & Leiden: Brill, 2005.

Lapidus, Ira, *A History of Islamic Societies*. Cambridge: Cambridge University Press, 1988.

Lemay, Richard. *Abu Mashar and Latin Aristotelianism in the Twelfth Century, The Recovery of Aristotle's Natural Philosophy through Iranian Astrology*. Beirut: American University of Beirut, 1962.

'Masha'alah', s.v. *Dictionary of Scientific Biography*.

McCluskey, Stephen C. *Astronomies and Cultures in Early Medieval Europe*. New York: Cambridge University Press, 1998.

Masood, Ehsan. *Science and Islam*. London: Icon Books, 2009.

'Muhyi al-Din al-Maghribi'. s.v. *Dictionary of Scientific Biography*.

Nasr, S. N. *An Introduction to Islamic Cosmological Doctrines: Conceptions of Nature and Methods for Study by the Ikhwan al-Safa, al-Biruni, and Ibn Sina*. Binghamton: State University of New York Press, 1993.

North, John David. *Cosmos: An Illustrated History of Astronomy and Cosmology*. Chicago: University of Chicago Press, 2008.

Park, Derek and Julia Park. *A History of Astrology*. London: André Deutsch, 1983.

Pingree, David. *From Astral Omens to Astrology: From Babylon to Bikaner.* Roma: Istitute italiano per l'Africa et l'Oriente, 1997.

———— *The Thousands of Abu Ma'shar.* London: The Warburg Institute, 1968.

Peterson, O. *A Survey of the Almagest.* Odensa, 1974.

Poulle, Emmanuel, 'The Alfonsine Tables and Alfonso X of Castile', *Journal of the History of Astronomy* 6 (1988): 85–112.

'Qadizade al Rumi: Salah al Din Musa ibn Muhammad al Rumi', in *The Biographical Encyclopedia of Astronomers.* Ed. Thomas Hockey et al. New York: Springer Reference, 2007.

'Qutb al Din al Shirazi', s.v. *Dictionary of Scientific Biography.*

Ragep, F. Jami. 'Tusi and Copernicus: The Earth's Motion in Context', *Science in Context* 14 (2001): 145–63.

———— *Nasir al Din Tusi's Memoir on Astronomy,* 2 vols. New York: Springer-Verlag, 1993.

———— 'Two Versions of the Tusi Couple', in *From Deferent to Equant.* New York, 1987. Pp. 329–56.

Richards, E. G. *Mapping Time: The Calendar and its History.* Oxford: Oxford University Press, 1999.

Robbins, Frank E. *Ptolemy's Tetrabiblos.* Cambridge, MA: Harvard University Press, 1940.

Sachau, E. (ed.). *Al-Biruni's India: An Account of the Religion, Philosophy, Literature, Geography, Chronology, Astronomy, Customs, Laws, and Astrology of India.* London: Keagan Paul, 1910.

———— (ed. & tr.). *The Chronology of Nations.* London, 1879.

Saliba, George. *A History of Arabic Astronomy: Planetary Theories During the Golden Age of Islam.* New York: New York University Press, 1994.

———— *Islamic Science and the Making of the European Renaissance.* London & Cambridge, MA: MIT Press, 2007.

Said-Azkhan, H. M. *Al-Biruni: His Times, Life, and Works.* Karachi: Hamdard Foundation, 1981.

Samso, Julio. *Islamic Astronomy and Medieval Spain.* Aldershot: Ashgate Publishing Ltd, 1994.

Sarma, Sreeramula Rajeswara. *Astronomical Instruments in the Rampur Raza Library.* Rampur: Rampur Raza Library, 2003.

Sayili, Aydin. *The Observatory in Islam and Its Place in the General History of the Observatory,* 2nd. edn. Ankara: Turk Tarih Kurumu Basimevi, 1988.

———— "Ala al-Din al Mansur's Poems on the Istanbul Observatory', in *Belleten* 20 (1956): 458–83.

Sharma, Nirendra Nath. *Sawai Jai Singh and his Astronomy.* Delhi: Motilal Banarsidass Publishers, 1995.

Shukla, Kripa Shankar. 'Main Characteristics and Achievements of Ancient Indian Astronomy in Historical Perspective', in *History of Oriental Astronomy: Proceedings of an Internattional Astronomical Union Collosquium,* no. 91. Ed. G. Swarup, A. K. Bag, K. S. Shukla. New Delhi, India, 13–16 November, 1985. Pp. 9–22;

Singh, Simon. *Big Bang: The Origin of the Universe*. New York: HarperCollins, 2005.

Swerdlow, N. M., 'Alfonsine Tables of Toledo and Later Alfonsine Tables', *Journal of the History of Astronomy* 25 (2004): 479–84.

'Tarikh', s.v. *Encyclopaedia of Islam*, 2nd edn.

Tekeli, Sevim. *The Clocks in Ottoman Empire in 16th Century and Taqi al-Din's 'The Brightest Stars for the Construction of the Mechanical Clocks'*. Ankara: T. C. Kültür Bakanliği, 2002.

———— 'Astronomical Instruments of Tycho Brahe and Taqi al-Din', in *Proceedings of The International Symposium on the Observatories in Islam (19–23 September 1977)*. Ed. Muammer Dizer. Istanbul: Milli Egitim Basimevi, 1980. Pp. 33–43.

Tirtha, S. G. *The Nectar of Grace: Omar Khayyam's Life and Works*. Allahabad, 1941.

'Tusi, Nasir al-Din'. s.v. *Encyclopaedia Iranica*.

'Ulugh Beg: Muhammad Taraghay ibn Shahrukh ibn Timur', in *The Biographical Encyclopedia of Astronomers*. Ed. Thomas Hockey et al. New York: Springer Reference, 2007.

Wickens, G. M. (tr.). *Akhlaq-i Nasiri*. London: George Allen & Unwin, 1964.

Winter, H. J. J. and W. Arafat. 'The Algebra of Omar Khayyami', *Journal of the Royal Asiatic Society of Bengal, Sci.* 16 (1950): 27–77.

Yamamoto, Keiji and Charles Burnett (eds & trs). *Abu Ma'sar on Historical Astrology*, 2 vols. Leiden: Brill, 2000.

Index